CENTRAL POWERS SMALL ARMS OF WORLD WAR ONE

John Walter

The Crowood Press

First published in 1999 by
The Crowood Press Ltd
Ramsbury, Marlborough
Wiltshire SN8 2HR

British Library Cataloguing-in-Publication Data
A catalogue record for this book is available from the British Library.

ISBN 1 86126 124 1

Acknowledgements

Preparation of this book has been greatly assisted by Ian Hogg and Anthony Carter, who each supplied information, illustrations and advice above and beyond the call of duty. I am also grateful for the help of Joachim Görtz of Munich, for tracking down many illustrations of the 'guns in use'; to Joseph J. Schroeder of Glenview, Illinois; to Wolfgang Seel, whose contributions to the *Deutsches Waffen Journal* changed my view of the relationship between Mauser and the Turkish authorities; to Dr Rolf Gminder of Heilbronn, for access to his collection of Parabellum pistols; to Hans-Bert Lockhoven, for allowing me to use illustrations from his *Waffen-archiv–Archives d'Armes–Arms Archives* series; to Christian Cranmer and the staff of *Firepower International*, South Godstone, for their patience and hospitality; to Weller & Dufty Ltd of Birmingham for supplying information; to Gordon Hughes of Brighton, for access to several important guns and accessories; to David Penn of the Imperial War Museum, London; to Herbert Woodend of the Pattern Room Collection, Royal Ordnance plc, Nottingham; and to the many friends and correspondents who supplied details ranging from a single unit marking to photographs. The source of each illustration is credited individually, if appropriate, in the accompanying caption.

When all is said and done, however, the responsibility for this book remains with me. Close scrutiny and the passage of time usually reveal that work of this type is never entirely free of errors, but I hope that these have at least been kept to a minimum and that *Central Powers' Small Arms of World War One* provides an interesting read.

John Walter
Hove, 1999

Typeface used: Times.

Typeset and designed by D & N Publishing
Membury Business Park, Lambourn Woodlands
Hungerford, Berkshire.

Printed and bound by The Bath Press.

Contents

Introduction

When Gavrilo Princip aimed his Browning pistol at the heir to the Austro-Hungarian throne and pulled the trigger, little did he know that the shots would account not only for Franz Ferdinand and his consort, but ultimately also for the millions of men, women and children who died in the war of 1914–18. Cemented together by royal bloodlines and traditional alliances, Great Powers fought a Great War with weapons whose destructive capacity contrasted with medical care harking back to the days of the American Civil War.

The 'war to end wars' was often paradoxical. The heart-stopping prose and poetry it engendered had to be set against sickening mutilation, and genocide co-existed uneasily with a quaint – almost mediaeval – chivalry which lasted longest in the air. Pilots jousted to an aerial death in their gaily painted craft, neither asking nor giving quarter: below them, technology simply ground humanity inexorably into the mud, snow, sand and blood of the battlefield.

Aeroplanes and airships sought mastery in height and speed; great guns hurled shells at each other, unseen; poison gas brought a fearsome new dimension to the land war; and the stealth of the submarine challenged not only the mightiest battleship but also the ability of nations to feed themselves. Even civilians could be threatened with death hundreds of miles from the front line.

Compared with the aeroplane, the big gun, the submarine and the tank, the personal firearms of the infantrymen – efficient though they often were – had little strategic significance. Storm-troop tactics did have an effect on the development of light automatic weapons towards the end of the war, but radical advances in weapons were customarily overlooked in case they disrupted large-scale production of established designs.

A major lesson of World War One, implicit in the American Civil War but ignored thereafter, was that many simple guns were better than a few sophisticated ones. So great was the carnage in 1914–18, however, that evidence of defects in small-arms design was too often buried in No Man's Land and little was ever done to change the basic infantry weapons. The withdrawal of the Canadian Ross rifle, which was potentially very dangerous to its firers, was a major exception.

None of the protagonists could make regulation weapons in sufficient numbers to make good the unparalleled, but largely unforeseen losses. Production of Berthier, Enfield, Lee-Enfield, Mannlicher, Mauser and Mosin-Nagant rifles each ran into millions by the end of the war, yet shortages of front-line equipment were still common. These shortfalls were overcome only by impressing into emergency service virtually anything which could shoot, and by turning huge quantities of captured rifles against their former owners.

Central Powers' Small Arms of World War One could have been filled with campaign histories or personal reminiscences, but neither approach would have been ideal; few battles had much effect on small-arms design, and, though the infantry rifle was often the soldier's best friend, recollections understandably linger on other things. Emphasis, therefore, has been laid on the identification of the individual guns and an inkling, if no more, of how they worked.

Some of the guns were great successes. The German Maxims wreaked havoc wherever war was pursued. Dug-in on the Western Front, they were largely responsible for the terrible toll paid by the Anglo-French forces on the Somme in the summer of 1916; in the East, they cut down Russians in swathes; and in the South, given to the Turks, they reduced Allied expectations of Gallipoli from a rapid victory to mere survival.

However, most of the remaining German and Austro-Hungarian weapons – even the cumbersome Mauser infantry rifles – proved to be just workmanlike rather than outstanding. Mannlicher rifles, Schwarzlose machine-guns and Parabellum (Luger) pistols were inferior to their Allied equivalents, and historians still argue whether the Bergmann submachine-gun had even been issued in quantity by the time of the Armistice.

The Central Powers did not have a genius in the mould of John Browning, who combined a flair for design with an appreciation of the value of simplicity. Mannlicher had died in 1904, and though Paul Mauser lived until May 1914, it is arguable how much credit for the Mauser rifles and pistols is due to him and how much to the engineers employed by Waffenfabrik Mauser AG. The great power wielded by leading manufacturers also inhibited the work of lesser lights merely by controlling mass production. Though there was very little wrong with the Maxim, the standing of DWM – particularly in the export markets – undoubtedly inhibited the progress of the Bergmann machine-gun.

The most interesting guns of World War One, of course, were the experimental prototypes and the many old weapons that were impressed into military service. Few of these ever fired shots in anger, as they were customarily confined to garrison, training or lines-of-communication defence. But the Germans and (to a lesser extent) the Austro-Hungarians showed great ingenuity. Some guns were converted to fire different cartridges, whilst others received bayonets which were being made in foundries with no prior experience of weapons.

The most impressive aspect of small-arms production in World War One was the way in which tens of thousands of small manufacturers were co-opted to take part in sub-contract work. Teething troubles were both inevitable and substantial: the German 'Stern Gewehre' were impossible to assemble without the services of trained armourers, and the British Peddled Rifle Scheme was little better. Gradually, however, the idea of sub-contracted mass production, assembled centrally with fully interchangeable parts, began to gain credence.

As the war became increasingly protracted, and the fighting became more desperate, designers were given a chance to promote mould-breaking ideas. Few of these were ever perfected, but among them were some remarkable attempts: the mechanically-driven Gebauer machine-gun, for example, or the 2cm Becker cannon. The Becker left a very effective and outstandingly successful post-war legacy in the form of the Oerlikon. The submachine-gun also owes its popularity to World War One, and the first steps were taken towards the development of the assault rifle.

More progress is made in a year of war than in a decade of peace, and many lessons learned in 1914–18 were put to good use in the years that followed. But were the gains *really* worth such a dreadful cost …?

CONTENT

Central Powers' Small Arms of World War One is split into three. **Part One** gives brief details of the pre-1914 ordnance history of each of the participants, which is particularly relevant as many of the old single-shot weapons were dragged out of mothballs. Photographic legacies confirm that these guns did indeed serve, and that the Martini-Enfield and Gras rifles on the Allied side were matched by single-shot Mausers and Werndls wielded by men of the Central Powers.

Part Two provides a register of the regulation weapons which served each army on the outbreak of war in 1914. They are grouped in declining order of efficiency, starting with the front-rank guns and ending with obsolete patterns which were often being held in store.

Part Three presents a year-by-year guide to the changes made in the small-arms inventories during the war, when vast numbers of impressed and captured weapons were used alongside the regulation designs.

Pictured somewhere in Germany in 1915, two of these three infantrymen – apparently from Infenterie-Regiment Nr 75 – are armed with Gew. 88s, the clip-ejection port on the base of the magazines being clear. The centre man carries a Gew. 98.

PART ONE: ORDNANCE HISTORY

1   Austria–Hungary

The rump of the once-powerful Holy Roman Empire, Austria–Hungary exerted its influence across much of central Europe when World War One began – from Bohemia in the north to Bosnia-Herzegovina in the south; and from the Tirol in the west to Bukovina and Transylvania in the east.

The Dual Monarchy had faced so many internal problems in the nineteenth century that an undercurrent of discontent was still evident in 1914. Indeed, the *kaiserlich und königlich Armée* (*k.u.k. Armée*: Imperial and Royal Army, otherwise known as the 'Common Army') was remarkable not for fighting bravely, but for fighting at all. Its unique ethnic mix raised problems faced by no other European army on such an epic scale.

Only about a quarter of the men were drawn from German-speaking areas whereas virtually half were Slavs. German was officially the language of command, but as nine languages and several dialects were commonplace in the Dual Monarchy, instruction was usually given in the native tongue of an individual unit. Mixed forces adopted the language of the majority; and any group numbering greater than 20 per cent of the total strength could demand instructions of its own.

The potential chaos that lay in the various languages was reduced by grouping men ethnically, then garrisoning them far away from their home territory to prevent independence movements putting pressure on individual soldiers. Helped by the development of an unofficial language known as 'Army Slav', with a basic vocabulary of eighty key words, the policy worked remarkably well.

This cosmopolitan background often forged closer bonds between officers and men than in the German Army – where the gulf between the officer class and the rank-and-file was sharply drawn – but encouraged revolutionary fervour in 1918 to a much greater degree. Declarations of independence by both Czechoslovakia and Hungary forced the Austrians to capitulate on 4 November 1918, and the Dual Monarchy, having lost 4.82 million dead and wounded, ceased to exist.

Unlike Prussia of the early 1860s, which was still technologically backward, Austria had not only embraced industrial revolution comparatively early, but had also nurtured long-established gun-making traditions.

The principal manufacturing centre was the small town of Ferlach, where 50,000 Lorenz rifle-muskets had been made for the Austrian Army prior to 1866. However, Ferlach had never recovered from economic depression at the end of the Napoleonic Wars and had survived only by making inexpensive 'Trade Guns' for export to Africa and the Far East.

Some of the best-known gunmakers – understandably specializing in high-class sporting guns – worked in the imperial capital, Vienna. Greater inventiveness was to be found in the northern province of Bohemia, where a group of gunsmiths worked in Prague and another group in Weipert.

The most important agency, however, had been founded in 1834 in the village of Oberletten an der Steyr in the Oberdonau region. The founder, Leopold Werndl, had built up a flourishing business by the 1840s, making components for firearms and edged weapons. As many as 500 workers were employed in workshops in Oberletten and Steyr.

An Austrian soldier of the 1860s with a Lorenz rifle musket, precursor of the Wänzl.

When Werndl died in 1855, he was succeeded by his son Josef (1831–89). The younger Werndl enlarged the factory, installing a steam engine to replace the capricious water-powered machinery, and soon realized that mechanization was the key to lasting success. In company with his factory superintendent, Carl Holub, Josef Werndl visited the USA at the end of the American Civil War to see at first hand how industry there had coped with unprecedented demand for firearms, and how the design of firearms had evolved to meet the need for increased firepower.

Inspired by the American experience, Werndl and Holub arrived back in Austria determined to make guns that were completely interchangeable. Machinery was ordered from Pratt & Whitney in the USA and Greenwood & Batley in Britain. A change from making small parts for guns to making guns in their entirety was risky, but the gamble proved worthwhile: the Steyr factory employed 3,000 people by 1866. When work on the Werndl rifle began in earnest in 1867, weekly production capacity was rated at 5,000 guns.

The changes could not have come at a better time. Austria and Prussia, grudging allies in the war against Denmark in 1864, were soon at each other's throats in the Seven Weeks War of 1866. Commentators throughout Europe predicted an easy victory for the well-trained men of the Austrian alliance at the expense of the mixture of regulars and conscripts fielded by the Prussians. However, they had failed to predict the Austrians' penchant for suicidal bayonet charges *en masse* against better-positioned opponents; to appreciate the value of even the imperfect Dreyse breech-loading needle rifle; and to see the lessons implicit in the ability of the Prussians, though giving the Austrians several weeks' start, to mobilize before their opponents were even half-ready.

The culmination of a series of disastrous defeats, most notably at Königgrätz, was a humiliating peace treaty signed in Prague on 23 August 1866. Ironically, not only had Josef Werndl already laid the basis for series production of modern weapons in Austria, but some army officers – notably those who had been attached to Prussian units during the fighting against the Danes – had also realized the advantage of the breech-loading rifle. They had even managed to get the *Hinterladungs-Gewehr-Commission* (Breech-Loading Rifle Commission) formed in June 1864 to oversee trials.

Unfortunately, the Austrian Army was as hidebound in tradition as its most reactionary contemporaries; breech-loading was seen simply as a waste of development time, production facilities and ammunition. A needle-fire rifle-musket conversion submitted by Edward Lindner of New York had even been tested during a cursory examination of breech-loading systems only a few weeks before the Seven Weeks War began, but little had been achieved.

The crushing defeats at the hands of the Prussians in general, and the imbalance of the casualty figures in particular, changed the situation almost overnight. Seeking a scapegoat, whilst themselves avoiding the blame, Austria's ordnance authorities identified the single-shot Lorenz rifle-muskets as a root cause of defeat. The Dreyse breech-loading needle rifle was elevated overnight from pariah to super gun, and the quest for a suitable breech-loader began. Work began in earnest when the *Hinterladungs-Gewehr-Commission* was reactivated in September 1866.

RIFLE-MUSKET CONVERSIONS

Submissions included cap-locks and metallic-cartridge rifles, but the testers had sensibly decided that metal-case ammunition was preferable. After extensive trials with Albini-Braendlin, Friedrich, Lindner, Prasch and Snider designs, the *Hinterladungs-*

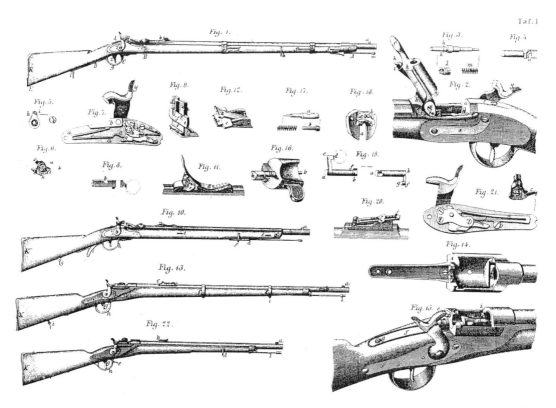

Drawings of the Wänzl and Werndl actions, from Kropatchek's Handbuch für die kais. kön. *Artillerie* (Vienna, 1873)

Gewehr-Commission recommended a swinging block breech-loading transformation developed by Franz Wänzl, a gunsmith trading in St Margarethen near Vienna. This had the important advantage of being indigenous, though essentially similar to many mid-1860s designs.

A few hundred Lorenz rifle-muskets were converted experimentally to the Wänzl breech in the winter of 1866. Field trials failed to find any weaknesses in the design, teething troubles were speedily overcome, and the Wänzl system was adopted officially on 5 January 1867.

The Model 1867 Wänzl rifle, often known as the M1863/67, was made on the basis of the cap-lock Lorenz rifle-musket. Wrought iron or steel barrels were used, depending on the manufacturing pattern (1854 or 1862 respectively); 1862-type guns also had smaller lock plates.

Converted by gunmaking establishments in Ferlach, as well as in Werndl's factory in Steyr, the Model 1867 Wänzl chambered a 13.9 × 33mm rim-fire cartridge which was originally defined as '6⅓ *linie*' or '6 *linie* 4 *punkt*''. The gun was loaded by thumbing back the hammer to half cock, raising the breech-block, inserting a new cartridge, and pressing the cartridge forward into the chamber. The block was shut, the hammer was retracted to full cock, and the trigger could be pressed. As the hammer flew forward to hit the striker, it also propelled a sliding bar into the back face of the breech-block, locking the mechanism shut at the moment of discharge

The rifles were 1,340mm long and weighed 4.25kg. The 885mm barrel was rifled with four grooves, each 5.3mm wide and 0.17mm deep, making a turn in 211cm. Muzzle velocity was about 407m/sec with a 26.4g bullet. Attaching the 1854-pattern bayonet, with a distinctive diagonal slot and the locking ring around the mid-point of the socket, increased length to 1,815mm and weight to about 4.65kg. The ramp-and-leaf sight was graduated to 1,100 *schritt* (845m) (= paces, *see* p. 61).

The Model 1867 Wänzl had a straight-wrist one-piece stock, with two spring-retained bands and a heavy nose-cap. A cleaning rod was carried beneath the barrel, whilst one swivel lay on the middle band – unusually close to the nose-cap – and another appeared under the butt.

Wänzl conversions were also applied to old *Jäger-Stutzen*, creating the M1867 *Stutzer* (short rifle). These had heavy octagonal barrels, with the muzzle crowns turned down to accept a sword-bladed bayonet. A locking ring was positioned around the base of the socket.

The *Ordinäre Stutzer* of 1853 had originally been rifled with four grooves and sighted to 1,000 *schritt* (770m), while the pillar-breech *Dornstutzer* of 1854 once had a heavy ramrod and sights graduated to 1,200 *schritt* (925m). The guns all had key-retained barrels, with swivels through the fore-end – above the cleaning-rod pipe – and on the under-edge of the butt. Trigger-guards ended in a finger-spur, while the backsights were distinctive curved-leaf 'grasshopper' patterns.

Survivors had been converted to fire expanding-ball ammunition after 1863, by which time they had been assimilated in a single group, but virtually all of those remaining in service in 1867 were speedily converted to breech-loading. A typical *Stutzer* M1853/63/67 had an overall length of 1,102mm, weighed 4.4kg, and had a 648mm barrel. The 1853-pattern socket bayonet increased length to 1,700mm and weight to about 5.1kg. The four-groove rifling made a turn in only 1,575mm, which made it appreciably 'quicker' than the rifle-musket pattern.

The Extra-Corps-*Gewehre* M1854/67 and M1862/67 were Wänzl conversions of the cap-locks of 1854 (iron barrel, large lock plate) and 1862 (steel barrel, small lock plate). Used by gendarmerie, sappers, pioneers and ancillary troops, they had a band in addition to a nose-cap, and accepted the M1854 socket bayonet. A typical M1862/67 example was 1,052mm long, had a 665mm barrel, and weighed 3.75kg without its bayonet. The ramp-and-leaf backsight was graduated to 500 *schritt* (385m). Attaching the bayonet increased overall length to 1,535mm and weight to 4.15kg. The pitch of the rifling duplicated that of the full-length guns: one turn in 2,110mm.

THE FIRST NEW
BREECH-LOADERS

Trials to find a brand-new breech-loading rifle were much more important than the concurrent search for a Lorenz rifle-musket transformation. The *Hinterladungs-Gewehr-Commission* demanded that muzzle velocity should exceed 340m/sec, but more than 100 designs were considered before the Commission selected the Austrian Würzinger, the American Peabody and the American Remington as the most desirable.

Protracted trials undertaken at the end of September 1866 eliminated the Wurzinger, and, when the Peabody was declared to be too difficult to load and clean, only the Remington remained. On 29 November 1866, the Commission recommended the purchase for troop trials of Remington rifles chambered for a rim-fire cartridge developed by the government laboratory. Several hundred Remington guns were then ordered from the USA.

However, after the competition had begun, a rifle designed by Josef Werndl and Carl Holub had been submitted to the *Kriegsministerium* (War Ministry). Tests undertaken secretly had shown that the Werndl-Holub breech mechanism was sturdy enough to withstand active service. One prototype had fired more than 2,000 shots over a ten-day period without serious problems, minor jams and failures to extract being blamed on the experimental cartridges.

The advent of the Werndl-Holub, which had not been entered in the official elimination trials, put the authorities in a potentially embarrassing position. The recommendation of the Remington was promptly undermined by the circulation of rumours that the rolling-block was too expensive, unsafe and that guns had exploded during trials. There is no evidence that this was true; though the original 'split-breech' Geiger-patent Remington was notoriously weak, the improved Rider-patent pattern tested in Austria was much more effectual.

Malicious rumour cleared the way for the adoption of the Werndl-Holub rifle on 28 July 1867 as the *Infanterie-und-Jägergewehr Modell* 1867. A stupendous order for 611,000 rifles was given to Josef Werndl, and the prosperity of his Steyr factory was assured.

THE WERNDL IN SERVICE

Problems with the mass-produced Werndl rifle became evident as soon as it entered service. Though the rifle was reasonably sturdy, and comparatively easy to use once the awkward breech-opening action had been mastered, it was also prone to jamming. Poor extraction was partly due to the drawbacks of the Wildburger-pattern cartridges, which had a badly designed case-head and an unnecessarily large primer which the firing pin in the breech-block customarily struck off-centre. The case body was apt to crack and, even though the slow-burning powder had been deliberately chosen to restrict the build up of pressure, the head was prone to fail.

These shortcomings were cured partly by improving the standard of manufacture, which reduced case-head failures appreciably, but mainly by introducing a stronger Roth-pattern case with a greatly improved head.

Though it looked simple, the Werndl breech mechanism contained a surprising number of pins and springs. Prolonged service showed that the breech-shoe or receiver was not as strong as it could be, and that the leaf springs enclosed in the drum unit soon became brittle. The backsight leaf was too weak, and few people liked the clumsy external hammer.

By far the worst problem of the Werndl, however, lay in its inability to accept a magazine – a fate that also overtook better designs such as the Remington and the Martini-Henry. Though many inventors developed quickloaders and magazine boxes, the Werndl defeated not only all of them, but also a radical auto-cocking adaptation proposed by Sylvestr and Karel Krnka.

Wracked by financial worries and internal discontent, the 'Dual Monarchy' of Austria–Hungary could rarely spare anything in the military budget to make more than token gestures in the hunt for a better military rifle until the mid-1880s. But changes were made in the basic drum-breech design.

The 1873-pattern Werndl rifle had a stronger receiver than its predecessors, weight being saved by paring away parts of the drum. Coil-springs replaced the previous leaves; the backsight base was reinforced; the hammer was mounted inside the lock plate, much more centrally than on the 1867 pattern; and the power of the standard cartridge was increased while trials were made with alternative ammunition.

A much-modified cartridge was adopted in December 1878, though the rifles and carbines remained unchanged except for chambering and sighting arrangements. Better propellant appeared in 1881, increasing velocity and reducing trajectory height, but experiments with the earliest magazine rifles had already exposed the shortcomings of the Werndl breech.

Werndl rifles were confined to Austria–Hungary, except for small-scale deliveries to Persia (23,000) and Montenegro (20,000). After the introduction of Mannlichers in the 1880s, Werndls were passed down through second-line and lines-of-communication troops to the *Landwehr* and then into store. Surprising quantities of surviving Werndls were subsequently reissued during World War One for guard duties and home defence.

MAGAZINE RIFLES

Shortly after the first Werndls had been introduced, a short rifle developed by Ferdinand Fruwirth of Vienna appeared. Though the tubular magazine beneath the barrel was clearly derived from the Swiss Vetterli, then brand-new, the Fruwirth carbine had sufficient merit to interest the Austrian authorities.

The first examples of this carbine were issued for trials in 1869, possibly chambered for the standard 11×36mm rimmed cartridge, and the Fruwirth was formally adopted for the *Gendarmerie der Österreichischen Reichshälfle* (sometimes known as the *Cisleithanischen Gendarmerie*) and the *Serecaner-Corps*) on 23 May 1872. Issue was subsequently extended in 1874 to the mounted gendarmerie in the Tirol and Dalmatia.

The bolt-handle rib sufficed as a lock; no ejector was fitted; and the cocking-piece had a prominent spur. The straight-wrist stock had a trigger-guard with a spurred rearward extension or '*Jäger* grip', similar to that of the Werndl carbine of the day. There was a small nose-cap, a swivel lay on the under-edge of the butt, and a sling-loop was anchored laterally through the fore-end.

A few examples of the 1872-type carbine were made in Fruwirth's workshops, but the majority of the 12,000 produced were made by Österreich-ische Waffenfabriks-Gesellschaft of Steyr in 1872–75. Chambered for the 11×36mm cartridge,

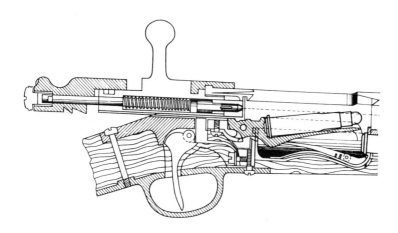

The action of the Kropatschek rifle.
Jaroslav Lugs

One of the earliest box-magazine Mannlichers, 1881. Konrad von Kromar

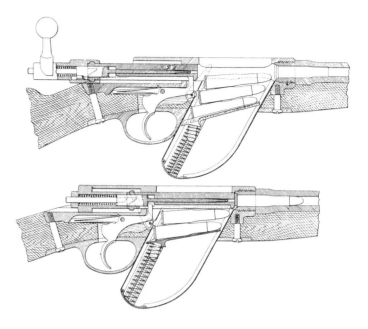

they were 1,035mm overall and weighed 3.15kg empty. The 560mm barrel had standard Werndl-pattern six-groove rifling. The tube magazine held six rounds, though a seventh could be placed on the elevator and an eighth in the chamber. The ramp-and-leaf backsight was graduated to 600 *schritt* (460m) and the *Stichbajonett für Extra-Corps-Gewehr* M1873 could be mounted, increasing overall length to 1,560mm and weight to 3.7kg. Muzzle velocity was about 305m/sec with the standard M1867 carbine round.

The Army tested the Fruwirth carbine in 1873 as a potential *Extra-Corps-Gewehr*. The gun was popular in the gendarmerie, owing to its handiness and good rate of fire, but proved to be too fragile to withstand military service; the locking system was only just safe enough for the cartridge. Surviving examples in service with the gendarmerie had all been withdrawn into store by 1890.

The failure of the Fruwirth carbine to interest the Austro-Hungarian Army cleared the way for a rifle patented on 2 November 1874 by Alfred von Kropatschek (1838–1911). Prototypes had been submitted to the Minister of War on 24 September 1874, with a conventional action inspired by the

German Mauser – using the bolt rib to lock against the receiver – and a Vetterli-type tube magazine in the fore-end.

By 1876, the Kropatschek was being declared as 'suitable for adoption', which, even though a Kropatschek gendarmerie carbine had been adopted in 1881, simply allowed a lengthy programme of minor improvements to drag onward. Leopold Gasser of Vienna patented a spring-loaded cartridge gate in 1879, by adapting existing Winchester patterns, and added it to the 'Gasser-Kropatschek' rifles tested in Austria–Hungary in the early 1880s. However, the rise of the box magazine restricted distribution – except in France, where the popularity of the Kropatschek in Indo-China and Equatorial Africa eventually created the Lebel.

Trials with the Kropatschek rifle dragged on for so long that the basic design had been overtaken before it could be properly exploited. So many guns had been offered for trial by the early 1880s that the Austro-Hungarian authorities, realizing that magazine-feed guns were being adopted by their neighbours (and potential foes), undertook trials with greater purpose than at any time in the preceding decade.

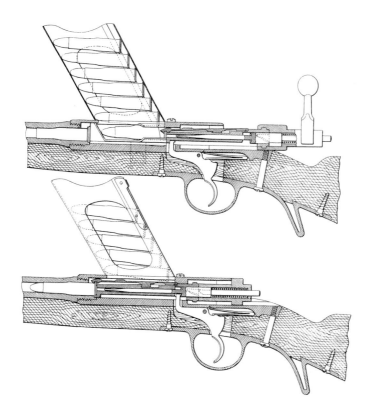

A Mannlicher rifle with a gravity-feed magazine, 1882. Konrad von Kromar

Among the most important submissions were those of Ferdinand Mannlicher (1848–1904), trained as a railway engineer and later ennobled, who had turned to small arms design in the late 1870s. The earliest rifles were unsuccessful – four different patterns were tested in Britain in the early 1880s, and all four failed.

Few of the trials rifles were efficient enough to catch the eye of the testers until the first of the straight-pull Mannlicher bolt systems (*Geradzug-Verschluss*) was tried. This had two lugs on a locking piece that connected the *Kammer* with the *Griffstück*. Fed from a five-round gravity magazine offset on the left side of the receiver, the 11mm-calibre rifle had a one-piece straight-wrist stock with a single band and a nose-cap accepting a sword bayonet. Swivels lay beneath the butt and band.

However, Mannlicher soon abandoned this gun for the simplified prototype of 1885. Made by

Österreichische Waffenfabriks-Gesellschaft and chambered for the 11mm, necked Werndl cartridge, the *Repetier-Gewehr 'Österreichische Vorlage'* relied on a bar under the back of the bolt being pivoted down into the bolt-way floor as the bolt was closed. The action was much simpler to make than the complex helical-channel bolt head of the earlier guns, and the clip-loading system was a great step forward. Empty clips were ejected through the top of the open action after the last rounds had been chambered.

The Mannlicher rifles were 1,328mm long, had 808mm barrels, and weighed about 4.75kg empty. They had standard six-groove rifling, and the ramp-and-leaf sights were graduated to 1,600 *schritt* (1,200m). The one-piece stock had a distinctively pointed pistol-grip. Two bands and a nose-cap were pinned in place; a lug for the special épée bayonet lay on the right side of the nose-cap; and a cleaning rod protruded beneath the muzzle. A large radial

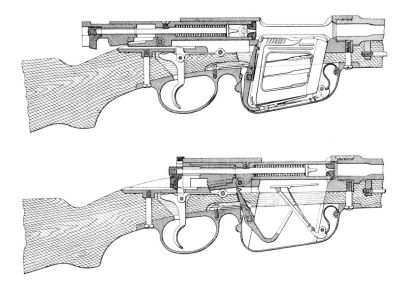

The Mannlicher M1885 'Österreichische Vorlage', precursor of the perfected design. Konrad von Kromar

catch on the right side of the magazine casing could be pressed to eject unwanted clips.

About 1,520 1885-pattern Mannlichers supplied by Österreichische Waffenfabriks-Gesellschaft were issued for field trials in the spring of 1886, passing the tests successfully enough to be recommended for service if minor improvements could be made.

The clip-release catch was greatly simplified and moved to the lower back of the magazine casing. The leaf-type backsight was replaced by a *Quadrantenvisier* (tangent sight); the stock was refined and lightened; the cleaning rod under the muzzle

was deleted; and the nose-cap was redesigned to accept a stacking rod. A short-bladed knife bayonet was substituted for the épée, and the rifle was officially adopted on 20 June 1886 as the *Infanterie-Repetier-Gewehr Modell* 1886.

THE SMOKELESS REVOLUTION

The adoption of the 1886-pattern Mannlicher rifle, chambering the Werndl cartridge, was a catastrophic error. Though it had ensured that wholesale

The M1886 with the action open, showing the locking wedge. Ian Hogg

15

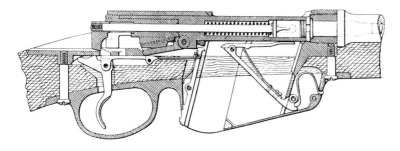

The Mannlicher M1888. Konrad von Kromar

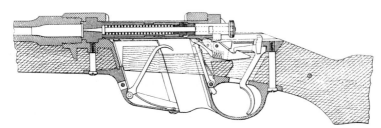

The Mannlicher M90. Konrad von Kromar

changes to ammunition supply were unnecessary – a cheap option! – concurrent trials had already shown that a calibre of 8mm had more to offer than the cumbersome 11mm.

When the French introduced a small-calibre cartridge loaded with smokeless propellant, just as the Austro-Hungarians began to issue the M1886, searching questions were bound to be asked. Blame fell on the Inspector of Infantry, Archduke Rudolf, a convenient scapegoat who subsequently committed suicide – contrite not for adopting the black-powder Mannlicher, but instead for political indiscretions and an ill-starred affair with a teenage baroness.

Even though his straight-pull bolt action had been adopted for military service, Ferdinand von Mannlicher continued to prepare new designs. The 1887/88 pattern was the first of them to embody a spool magazine patented by Otto Schönauer, and had a straight-pull bolt system with the locking lugs on the connecting-piece between the bolt head and the body.

The eight-round magazine, which had a single-finger follower, could be loaded from a special *Patronen-Packet* (stripper-clip). The one-piece stock had a pistol-grip butt, there were two screwed bands, and the nose-cap had a stacking rod and a bayonet lug. Swivels lay beneath the butt and the middle band, and a quadrant backsight appeared on the barrel. Trials revealed the rifle to be too complicated for arduous service, and it was rejected in favour of the M1888 Mannlicher.

The Mannlicher M95. Konrad von Kromar

This also proved to be a mistake, as Mannlicher submitted the first of his perfected straight-pull designs in 1889. This consisted of a twin-lug bolt which rotated within the bolt sleeve when the operating handle was pulled back. Rotation was due to cam-tracks on the bolt acting in concert with lugs inside the sleeve to turn the locking lugs until they disengaged seats in the receiver immediately behind the chamber. Guns of this type were superior to the preceding wedge-lock Mannlichers, offering not only far greater strength but also a considerable reduction in length.

The 1890-pattern short rifles and carbines were the first representatives to be accepted for service, though substantial numbers of similar infantry rifles were issued for troop trials. These dragged on until the perfected *Repetiergewehr* M1895 was approved in 1896.

The great success of the Mannlicher rifles had a beneficial effect on the prosperity of the Steyr factory which, by the early 1890s, had risen to become one of the most important small-arms manufacturing centres in Europe. The annual report for 1889–90 recorded that 469,070 of the 760,631 rifles on the order books had been delivered during the financial year, and that the receipt of additional orders promised full employment in 1890–91. Weekly output during this period was rated at 8,000–11,000 rifles.

AUTOMATIC WEAPONS

Impressed by a trial in Vienna in 1888, the Austrians purchased 160 Maxim machine-guns from Britain in 1889. Most of these were apparently intended for fortress use, but a few were given to the Austrian Navy. The Maxims were chambered for the 8mm cartridge and worked extremely well, but political pressure was brought to bear on the authorities – by no means uncommon in Austria–Hungary – to adopt an indigenous design.

The Škoda gun, promoted by the renowned arms-making company of the same name, was the work of two soldiers with influence at the highest level: Archduke Karl Salvator and Georg, Ritter von Dormus. Designed in 1885, only a year after Maxim had been granted his first patent, the Salvator-Dormus gun was seen as a cornerstone of the business formed by the German-trained engineer Emil Škoda (1839–1900).

Deported from Germany at the time of the Seven Weeks War in 1866, Škoda became works superintendent of an arms-making workshop founded in 1859 in Valdstejn by Graf Arnost. After purchasing the facilities in 1862, Škoda was putting the knowledge of steel-making he had gained in the Weser shipyard in Bremen to good use by the end of the decade. A steelworks was opened in 1884 and by 1890, the company was advertising a range of guns, gun carriages, ammunition, cast-steel armour plate and the '*Patent Mitrailleuse*'.

Experiments had been made with the Salvator-Dormus prototypes as early as 1886, after an agreement had been reached with the inventors. By 1889, the basic design had been finalized with the assistance of technicians led by Andreas Radovanovic; and, in accordance with the custom of the day, a patent was sought in Škoda's name.

Three Škoda guns, Nos. 3, 7 and 8, were tested over a three-year period by the *Militär-Technische-Komitee* (Military-Technical Committee). The guns proved to function reliably at elevations as great as 25 degrees and depressions of 30 degrees or more, though the fire-rate often had to be adjusted to allow for the angle of the barrel. The parts were easily exchanged, accuracy was acceptable, and durability was surprisingly good. Gun No. 8 eventually fired 40,787 rounds without any of the major parts breaking or wearing out.

In October 1893, therefore, the Škoda was accepted for service instead of the Maxim. How a preference for such a bizarrely constructed weapon could be expressed at the expense of the Maxim beggars belief. Nevertheless, the 1893-type Škoda remained in service in armoured positions until World War One.

The manufacturer endeavoured to obtain export orders, exhibiting guns in 6.5mm, 7mm and 8mm at the Paris Exposition in 1900; however, only 210 had been made by 1901. It is assumed that Austria–Hungary was using most of these, but the actual total is not known. A few guns, landed from warships, were used to defend the Legation in Peking during the Boxer Rebellion of 1899–1900. There they are said to have made a significant contribution to the fighting, but very little active service was ever seen elsewhere.

The Škoda had never been battle-worthy, and work on the M1893 stopped in favour of an improved design in 1901. The gravity-feed magazine and the pendulous rate-reducer were greatly refined, but the changes could not disguise the fact that the weapon was obsolete.

The Model 1903 was a long-barrel derivative of the M1901, mounted on a light tripod intended for cavalrymen, but only six were made; they were tested by the Austro-Hungarian authorities in 1905, on a light wheeled carriage, but were rejected in 1906 after innumerable problems had occurred. The 765mm barrel raised the muzzle velocity to 620m/sec; weight was about 24kg without the tripod mount (which added 7kg); rate of fire was listed as 300–600rd/min, owing to changes in the pendulum mechanism; and maximum range was claimed to be 4,500 *schritt* (3475m).

Though the first Austro-Hungarian machineguns had seen success, the same could not be said of the earliest auto-loading rifles. Introduced in 1885, the quirky recoil-operated Mannlicher *Handmitrailleuse* fed automatically from a ten-round gravity magazine above the left side of the breech.

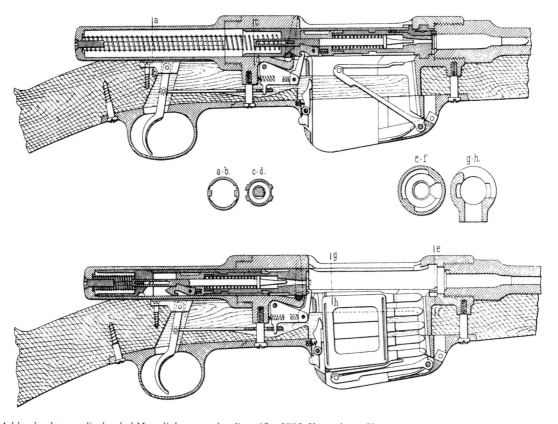

A blowback-type clip-loaded Mannlicher auto-loading rifle, 1893. Konrad von Kromar

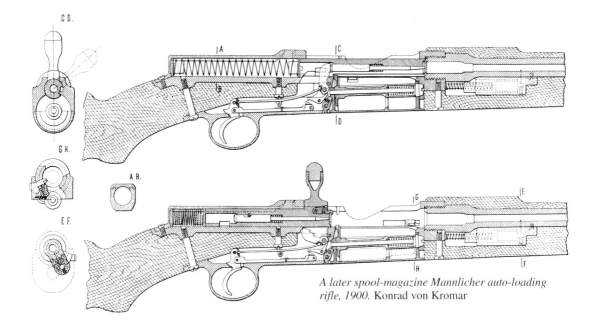

*A later spool-magazine Mannlicher auto-loading
rifle, 1900.* Konrad von Kromar

When the gun fired, the barrel recoiled until
struts or *zange* (tongs) dropped into a well in the
frame and allowed the breech-block to run back
against a powerful spring. The block then returned,
stripping a new round into the chamber from a carrier pivoted in the left wall of the receiver, and
raised the tongs into their upper position. The barrel and bolt then ran back into battery and recon-
nected the trigger system for the next shot.

Unfortunately, the *Handmitrailleuse* was a
clumsy weapon, with its wood half-stock, finger
grooves in the short fore-end, and a ladder-type
backsight folding down above the breech. It was
not successful.

The 1891-type rifle – an improved *Handmitrailleuse* – was locked by a transverse lug on the
top rear of the tongs engaging a recess on the
underside of the bolt. The lock was broken when a
shoulder in the frame, ahead of the fixed
clip-loaded magazine, cammed the tongs downward as the recoil stroke commenced. The rifle had
a short two-piece stock joined in front of the backsight. The sliding barrel was contained in a
sheet-metal jacket, and the nose-cap had a bayonet
lug on the right side. The sear, disconnector and

hold-open system were too complicated to be
durable, and, like its immediate predecessor, the
1891-pattern Mannlicher rifle was a failure.

1893 brought the first of the delayed blowback
Mannlichers to be extensively tested by the Austro-Hungarian Army as an infantry weapon. One
type relied on steeply pitched lugs on the bolt head
rotating against the resistance provided by the
seats in the receiver. The one-piece stock had a
straight-wrist butt, a single band, and a nose-cap
with a stacking rod and a lug for a sword bayonet.
A clip-loaded single-column magazine projected
beneath the stock, well ahead of the trigger-guard.

Another series of rifles was built on straight-pull
bolt systems, with the bolt head connected with the
body by cam-lugs moving in helical grooves. The
principal visible difference was the position of the
bolt-handle, which reciprocated in a straight path
on the right side of the breech and did not intrude
into the sight-line during the operating cycle as it
did in the turning-bolt design.

Neither pattern was successful, trials revealing
that they extracted too violently to function efficiently unless special ammunition and adequate
lubrication were used.

The M1895 gas-operated semi-automatic rifle, externally resembling a lever-action Winchester, had a two-piece stock and a full-length hand guard above the barrel. A five-round magazine was contained in the receiver. A charging handle on the right side of the receiver could be used to cock the hammer and load the chamber for the first shot.

When the gun fired, however, gas tapped from the bore forced the cocking slide back – cocking the hammer, pivoting the breech-block sideways to remove the lateral locking lug from its recess in the receiver behind the chamber, and operating the 'L'-shaped extractor on the right side of the breech. Though the action was extremely compact, it was unacceptable militarily.

By 1900, Mannlicher had developed the gas-operated design that provided the basis for guns perfected posthumously in 1905–08. Gas tapped from the bore ahead of the chamber forced an operating rod backward so that its tip, protruding through a slot in the left side of the receiver, could rotate the bolt about 45 degrees anti-clockwise to disengage the bolt-handle base from its seat in the receiver. The guns were hammer-fired, with detachable five-round spool magazines in a housing beneath the feed way.

Guides for stripper-clips were machined into the receiver. Pistol-grip butts were accompanied by separate fore-ends, with full-length hand guards and a single band, and a combined stacking rod/bayonet lug lay on the underside of the Swiss-style nose-cap.

Guns of this pattern initially suffered from a weak breech-lock, rapid erosion of the gas port, and the intrusion of the charging handle into the sight-line during the firing cycle. However, greatly improved forms were still being tested in Germany and Austria–Hungary as late as 1908 and the best of them proved to be efficient weapons.

HANDGUNS

The standard service weapons of the Austro-Hungarian Army prior to 1907 were revolvers derived from the original Gasser patterns. The oldest dated from 1870 and lasted in an improved form as late as World War One.

Though strenuous attempts were made to perfect the mechanical repeater in the last quarter of the nineteenth century – mainly in Bohemia – few had any lasting effect on technology. It is suspected that the Bittner pistol achieved the widest distribution, but was never successful militarily. However, the mechanical repeaters developed by Josef Laumann in the early 1890s led directly to the Schönberger pistol. Among the earliest autoloading handguns, though still the subject of controversy, this was tested by the military authorities *c*.1895.

The Schönberger pistol.

The 1894-pattern 'blow forward'
Mannlicher. Eidgenössische
Waffenfabrik, Bern

The operating system of the Schönberger pistol has been debated on many occasions, as some writers have identified it as a delayed blowback while others have claimed it to be an example of primer actuation. In addition, the date of introduction is also vigorously contested.

Though the Austro-Hungarian authorities were happy to test handguns such as the Schönberger, the Kromar and the Salvator-Dormus, few of the first generation of auto-loading pistols were efficient enough to interest them. With the exception of the Borchardt (patented in 1893) and the Mauser

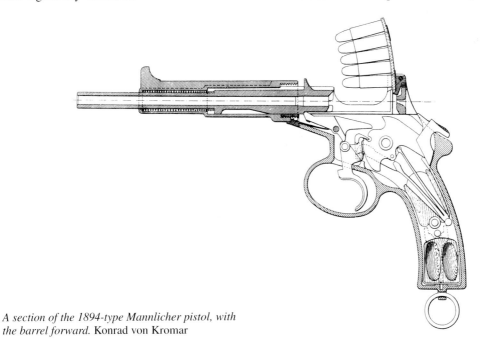

A section of the 1894-type Mannlicher pistol, with the barrel forward. Konrad von Kromar

(1895), few fired bullets that could rival revolver patterns for stopping power.

Better designs eventually prevailed. The first Mannlicher pistol, protected by a German patent granted in December 1894, had a barrel that was projected forward from the standing breech. This 'blow-forward' action – rarely encountered in firearms, and equally rarely successful – compressed the mainspring between the barrel and the front of the breech housing during the operating stroke. Semi-automatic operation was ensured by the double-action revolver-type trigger, which had to be released before another shot could be fired.

The Mannlicher design was simple, but had some poor features: the movement of the barrel made the guns difficult to fire accurately, the charger-loaded magazine was inferior to a detachable box, and the original rimmed ammunition promoted jamming. Most guns chambered a 7.6mm cartridge, though a few 6.5mm variants are known. Only about 200 M1894 Mannlicher pistols were made, mostly by Waffenfabrik Neuhausen (now SIG) in Switzerland.

An unusual blowback Mannlicher design, never exploited commercially, was patented in 1896. It relied on a special hammer, directly behind the magazine well, with a spur on an elongated cocking lever protruding at the rear of the frame. A second patent granted in 1896 protected a much better design. Commonly known as the Mannlicher M1896, the earliest examples of this handgun appeared in 1897 and were entered in the Swiss trials of 1898. Few had been made by 1900, but the basic design was subsequently revived as the Model 1903 and offered as a pistol or a pistol-carbine.

The breech mechanism was weak compared with that of rivals such as the Mauser C96 and the Borchardt-Luger. Although superficially resembling the Mauser externally, the 1896/1903 pattern Mannlicher differed greatly internally. The German design was locked by a sturdy rising block beneath the bolt, but the Mannlicher had a strut between the bolt and the rear of the receiver which was barely robust enough to handle even the low-powered variant of the 7.65mm Borchardt cartridge.

The perfected Mannlicher pistols were elegant, though comparatively weak blowbacks of conventional design. Protected by a German patent granted in October 1898, they were marketed as the models of 1900, 1901 and 1905. The earliest examples were made by Waffenfabrik von Dreyse in Sömmerda, but later versions were the work of Österreichische Waffenfabriks-Gesellschaft in Steyr (1901–05). They were successful only in Argentina, where the high price and considerable complexity of the Mauser and Borchardt-Luger allowed Österreichische Waffenfabriks-Gesellschaft to find a niche market for the Mannlichers.

The so-called 'M1896' Mannlicher was a recoil-operated design. Eidgenössische Waffenfabrik, Bern

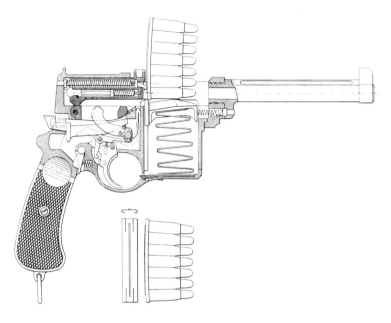

A section of the 1896-patent Mannlicher, with the charger in the loading position. Konrad von Kromar

Mannlichers rarely challenged the Borchardt-Luger if complexity and price were unimportant, and were neither as simple nor as sturdy as the perfected blowback FN-Brownings. The Germans allowed the Mannlichers to take part in military pistol trials only to test whether they were suitable to be bought privately by officers.

About 10,000 Mannlicher pistols (mostly Model 1905) were sold over a ten-year period. The designs were obsolete by the time World War One began, though some no doubt saw action as officers' weapons.

Revolvers continued to serve all sections of the Common Army until 1907, when the *Repetier-pistole* M7 or Roth-Steyr was adopted for the cavalry. Some features of this gun originated in a patent granted in 1895 to Wasa Theodorovic, but the basic recoil-operated locking mechanism is

A short-barrel 'M1905' Mannlicher: the perfected blowback design. Weller & Dufty Ltd

23

A typical early Roth-Steyr pistol, 1898. Eidgenössische Waffenfabrik, Bern

usually attributed to Karel Krnka. Krnka-Roth pistols were tested throughout Europe from 1898 onwards, until the perfected version was issued for field trials with the Common Army in 1905–06.

The M7 pistol was claimed to embody important advantages as a cavalry weapon, particularly in the isolation of the trigger system from the autoloading action to reduce the chances of firing accidentally. However, the guns were complicated, loaded from chargers, and were expensive to make. Though substantial quantities were made prior to World War One, in Steyr and Budapest, the

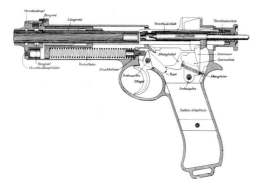

A sectional drawing of the Roth-Steyr, from the official handbook.

Repetierpistole M12 or Steyr-Hahn superseded them after 1914. This derived from the same basic recoil operated rotating-barrel action, but it had been greatly refined by Steyr technicians as a commercial venture. Apart from the charger-loaded magazine, the M12 was a much better combat weapon than the *Repetierpistole* M7.

However, although the Austro-Hungarian Army made good use of the *Repetierpistolen* M7 and M12, the second-line forces, the Austrian *Landwehr* and the Hungarian *Honvéd*, were equipped differently. The Austrians seem to have retained the M1898 Rast & Gasser revolver until World War One, but the Hungarians introduced a Frommer '12.M' pistol in 1912.

THE LATER ŠKODA MACHINE-GUNS

Tests against the Maxim and the Schwarzlose predictably revealed the shortcomings of the Škoda. However, though interest had waned in machine-gun design with the adoption of the Schwarzlose in 1907, Škoda continued development in the hope of attracting export orders. The basic block-type breech mechanism was surprisingly efficient once the

The commercial version of the Repetierpistole *M12 ('Steyr-Hahn').*

pendulum-type rate reducer – a needless complication – had been abandoned and a belt-feed system had replaced the unsatisfactory gravity-feed box.

The 1909-pattern Škoda machine-gun was being demonstrated by the summer of 1910, but the success of this particular gun is difficult to determine. Claims that it served the Austro-Hungarian Army in 1911–13, as an expedient pending delivery of M07/12 Schwarzlose guns, have proved difficult to verify.

Škoda advertised the M09 in 6.5mm, 7mm, 7.65mm, 7.9mm and 8mm (8 × 50R), and trials were undertaken for Bulgaria, China, the Netherlands, Peru, Romania and Turkey prior to 1914. Yet only the Chinese are known to have purchased the guns in any numbers, as part of a contract signed in 1910 for a selection of Škoda's guns and armour-plate. It is believed that fifty 7.9mm-calibre machine-guns were to be supplied, but that work was stopped by the revolution of 1911 that swept away the Manchu Dynasty; most of the weapons remained in store in Pilsen until World War One began. About thirty of them are said to have subsequently been issued for field service in Austria–Hungary, mostly aboard an armoured train.

The M13 and M14, which were identical apart from the date of manufacture, superseded the M09.

These guns had longer barrels, improved belt-feed systems, and more efficient lubricators. They were mounted on the M13 tripod, which could be distinguished from the 1909 pattern by its smaller size and by steel links connecting the legs to prevent collapse in the firing position. Production of Škoda machine-guns ceased when World War One began, their demise being partly due to the emergence of the Schwarzlose.

Andreas Schwarzlose (1867–1936), born in Germany, is one of the more interesting of the pre-1914 inventors. Unlike contemporaries such as Mauser, whose reputation depended largely on the perfection of a single design – and then often with the help of others – Schwarzlose was always prepared to try something different.

Testimony to this is provided by a range of pistols dating from the 1890s, culminating in a military pattern that was so nearly a great success. He also developed the only blow-forward design to find commercial favour, selling several thousand in Europe before the rights were hawked to North American interests prior to World War One.

The Schwarzlose machine-gun had been perfected by 1902. Compared with the recoil-operated Maxim, the delayed-blowback mechanism was extremely simple. It was easy to make and proved to be reliable once the rapid opening of the breech had been delayed sufficiently. The barrels of the earliest guns were too long, resulting in chamber jams and case-head separations, but the problem was cured simply by cutting down the 650mm barrel of the experimental 1905-type gun to 530mm on the perfected *Maschinengewehr* M07.

Made under licence by Österreichische Waffen-fabriks-Gesellschaft, the Schwarzlose M07 and M07/12 machine-guns not only survived until the

Two illustrations of the field-trials Schwarzlose machine-gun, from an Austro-Hungarian manual of 1908.

end of World War One, but also equipped the post-war Austrian, Hungarian, Czechoslovakian and Yugoslav armies. Large numbers had been made under licence in Sweden and the Netherlands even before World War One began.

DECLINE AND FALL

The 1890s represented the high point of the Austro-Hungarian small-arms industry. Österreichische Waffenfabriks-Gesellschaft had sold substantial quantities of *Reichsgewehr*-type rifles and carbines to Peru and Brazil, representing some of the first small-bore magazine rifles ever to see large-scale combat experience, and Mannlichers derived from the *Reichsgewehr* had been adopted by Romania in 1893.

Chile advertised its intention to hold trials in Paris in 1892, and the Steyr factory management, which had already obtained orders from Chile for Gras and Kropatschek rifles, was confident of success. However, Chile eventually decided to take a Spanish-type Mauser and it is probable that the aggressive support of the German government was a deciding factor.

Portugal proved another great disappointment. Though Mannlicher-Schönauer rifles were acquired in substantial quantities for trials in 1900–01,

Portugal elected to develop a gun of its own – the Verguiero-Mauser – and then placed the order with Deutsche Waffen- und Munitionsfabriken. By the early 1900s, therefore, the lackadaisical and often introspective attitude of the government of the Dual Monarchy was contrasting starkly with that of Germany, where the manufacture of weapons was seen as an important tool of trade.

Early in 1897, Österreichische Waffenfabriks-Gesellschaft had been forced into a German-controlled cartel to ensure a share of the production of Mauser rifles. Output of the Mannlichers was declining, as the initial Romanian orders had been fulfilled, the Dutch were making guns under licence, and the Austro-Hungarian government orders were not sufficient in themselves to keep the workforce fully employed. Yet, when World War One began, the Steyr factory had made the staggering total of 6,065,234 military rifles and carbines, 284,447 pistols, 9,215 machine-guns and about 20,000 sporting guns.

A new factory was built in Steyr in 1913–14, but the Austro-Hungarian government still placed nothing other than token orders. A great shortage of serviceable rifles became evident as soon as mobilization began in the summer of 1914, and was partly solved, ironically, by seizing 7mm-calibre Mausers being made in Steyr for Mexico.

Even large-scale arms makers were still primitively equipped prior to World War One. This 1908-vintage view of the Fabrique Nationale cartridge factory (part of the same cartel as Österreichische Waffenfabriks-Gesellschaft) gives a good idea of the working conditions.

2 Bulgaria

Bulgaria was ruled by Turkey until the Russo-Turkish War of 1877–78. The Treaty of San Stefano initially created 'Big Bulgaria', an independent state, but this was subsequently reduced by the Congress of Berlin (1879) to a small autonomous principality within the Ottoman Empire. Independence was declared unilaterally in 1908, when Ferdinand I took the title of king. Bulgaria was a founder member of the Balkan League, with Greece, Serbia and Montenegro.

Territory was gained as a result of the First Balkan War of 1912–13. Bulgaria then unwisely fought not only Turkey and Romania but also the other members of the Balkan League in the short Second Balkan War of 1913, losing much of the territory gained just a few months earlier.

Initially uncommitted in World War One, Bulgaria signed a secret pact with the Central Powers in July 1915, declaring war on Serbia in October in the hope of gaining territory lost in the Second Balkan War. Britain, France and Italy then declared war on Bulgaria a few days later.

Bulgarian troops played a minor role in the fighting, losing some 95,000 men dead and 155,000 wounded before an armistice was signed in Salonika on 28 September 1918; hostilities ceased two days later.

Lacking production facilities, Bulgaria's ordnance affairs were dictated largely by her neighbours. Though the initial influences were predominantly Turkish or Russian – Peabody-Martini and Berdan rifles were used for many years – these waned in favour first of Austria–Hungary (by 1890) and then ultimately of Germany. Appropriate marks will thus be found on Mannlicher rifles, Maxim machine-guns and Parabellum pistols. France financed a rearmament programme in 1906. Sensibly, the Bulgarians spent the money on good-quality Schneider artillery instead of the ineffectual French small arms.

3 Germany

The defeats inflicted on the Austrian Army by Prussia during the Seven Weeks War of 1866 – allies just two years earlier in the campaign against Denmark – were a pivotal point in European history. Austria's defeat marked the beginning of the decline of Habsburg power, and the emergence of a truly united Germany.

When the million-strong armies of Prussia, Baden, Bavaria, Hessen, Saxony and Württemberg went on to crush the French in the Franco-Prussian War of 1870–71, bolstered by the other members of the *Norddeutscher Bund* (North German Confederation), the seal was finally set on the *Deutsches Reich* (German Empire). The imperial constitution ratified in May 1871 gave much of the power to Prussia, beginning the march down a road of intrigue and suspicion that was destined to end in the misery of total war.

The unified post-1871 military structure allowed Bavaria considerable independence, but most of the other state forces were integrated in a single numbered series. Saxony retained its own Army List, controlling postings and promotion, whereas Baden, Hessen and Württemberg amalgamated their forces with those of Prussia.

The creation of efficient railways was a key to moving troops efficiently from one side of the *Deutsches Reich* to the other, and the development of a powerful arms industry was also given priority. Krupp was already making ordnance in quantity, but the emergence in the 1870s of Loewe, Mauser, the Lorenz ammunition business and propellant makers such as Rottweiler Pulverfabriken created the lucrative export business that ended only with the commencement of World War One.

The war against France in 1870–71 was fought largely with obsolete Dreyse *Zündnadelgewehre* (needle rifles), though the Bavarian units, controlled independently, were armed with a mixture of the Podewils-Lindner-Braunmuhl bolt-action breech-loaders (converted rifle-muskets) and pivoting-block Werder '*Blitzgewehre*'.

None of these weapons proved to be anything other than a developmental dead-end, for guns such as the Werder, whatever their merits, could not be converted satisfactorily to magazine feed. Transformations of rifle-muskets served only as a stop-gap until better designs could be introduced; and even though the bolt-action of the Dreyse rifle – thirty years old in 1870 – did indeed foretell the future, its needle-ignition system was too primitive to survive.

The immediate answer was found in a self-cocking adaptation of the Dreyse action developed in 1866 by Peter-Paul Mauser, a gunsmith employed in the Württemberg government armoury in Oberndorf. Speed of fire was greatly increased by eliminating the separate cocking movements. The needle sleeve of the Dreyse had to be retracted manually before the bolt could be opened for loading, but the Mauser version could be operated simply by raising the bolt-handle and pulling it backwards.

Peter-Paul and Wilhelm Mauser approached the authorities in Württemberg, but Minié-type expanding ball ammunition was so well established in the army that the self-cocking prototype was simply ignored. Though this seemed a great setback at the time, a far better design had soon been prepared.

Prussian infantrymen, carrying Dreyse needle guns, storm the heights of Roten Berg ('Red Hill') during the battle of Spicheren in the Franco-Prussian War.

THE FIRST MAUSER

By 1867, the Mauser brothers had developed a modified rifle chambering a self-contained metal-case cartridge. This was also shown to the army ordnance department, but a needle-fire conversion of the standard *Dorn-Gewehr* rifle-musket had just been adopted officially – possibly at Prussian insistence – and nothing more could be done. In desperation, the Mauser brothers approached the Prussian ambassador to Württemberg. Unfortunately, the victories gained in the Seven Weeks War, a recent and very vivid memory, had convinced even the most sceptical Prussian that the Dreyse was a wonder weapon. When the ambassador refused to listen, the Mausers turned instead to his Austrian counterpart in the hope that the drubbing suffered at the

muzzles of Prussian guns would make Austria more amenable to change.

The Austrian ambassador to Württemberg was keen enough to send the prototype Mauser to Vienna early in 1867, but a preference had already been expressed for the Remington Rolling Block. Even though the testers acknowledged that the Mauser had some very good features, the artillery committee had exhibited a liking for block-type actions.

Rejection of the prototype Mauser self-cocking bolt-action rifle by the Austrians was a grave setback. In the summer of 1867, however, General Arthur Graf Bylandt-Rheidt – president of the artillery committee and later Minister of War – mentioned the existence of the Mauser to Samuel Norris, the European sales agent of Eliphalet Remington & Sons.

Norris had contacts in most European ordnance circles, and immediately saw the potential in the Mauser prototype. He was also realistic enough to realize that the bolt-action system conflicted directly with the 'Rolling Block' Remingtons, and so his negotiations with the French military authorities were undertaken secretly.

The French soon realized that the Mauser breech was vastly superior not only to the Dreyse but also to their own Chassepot. Buoyed by the strength of feeling in French ordnance circles, Norris told the Mauser brothers of his plans and, by the summer of 1867, operations had moved to Liége to prepare a provisional patent for the 'Mauser-Norris' or 'C67/69' rifle.

S. NORRIS & W. & P. MAUSER.
Breech-Loading Fire-Arm.
No. 78,603.   Patented June 2, 1868.

Inventors:

A sheet of drawings from the patent granted to Samuel Norris and the Mauser brothers in June 1868. US Patent Office

A contract dated 28 September 1867 allowed Norris to control the exploitation of the rifle design at the cost of small sums of money spread over several years. Virtually the only clause in the Mauser brothers' favour was a guarantee that rights to the invention would revert to them if Norris failed to pay the annuities. In this guarantee was salvation.

THE END OF THE BEGINNING

Trouble started when Remington & Sons learned of the duplicitous dealings of their salesman. Though Norris claimed to have been acquiring rights to a competitor of the rolling-block rifle – and that Remington was to have made the Mauser-Norris if it proved to be successful – his paymasters were unimpressed. Norris was immediately fired.

The French government, the major source of interest in the Mauser breech system, was becoming so concerned about the imminence of war with Prussia that Chassepot needle-rifles were being issued as fast as possible; converting them to fire metal-case ammunition had become unthinkable. Interest in the Mauser project waned, and rights reverted to the Mauser brothers when Norris neglected to pay the first instalment of the third annuity in 1870.

Wilhelm Mauser soon proved to be not only an excellent salesman but also a skilled diplomat. He is believed to have made the first approaches to the Prussian Army in the spring of 1870, when an improved 'C70' rifle was tested by the *Militär-Schiess-Schule* (School of Marksmanship) in Spandau. The testers soon realized that the Mauser was ballistically superior to the Dreyse *Zündnadel-gewehre*, which were already being undermined by the Chassepot. The French needle-gun fired a smaller bullet at much higher velocity, reducing the height of trajectory and minimizing range-gauging errors, and though the Prussians were perfecting a conversion system – the Beck Transformation – the improvements still fell short of expectations.

Trials with the Mauser rifle began again in Spandau in the summer of 1870, but were interrupted by the Franco-Prussian War. By the early summer of 1871, the choice lay between the Mauser and the Bavarian Werder. Though each gun performed creditably, the Werder was complicated, difficult to keep clean, fouled excessively, and only marginally strong enough to withstand the overload-charge trials.

Ammunition presented a major problem. Though the Bavarians had been using metal-cased cartridges since the late 1860s, the reliance placed by the Prussians on the Dreyse needle-gun had left them with very little experience of modern metal-case cartridges and virtually no production capacity. Eventually, after protracted experimentation, an 11mm *Reichspatrone* was approved with a rimmed case and a reduced-diameter neck.

By 7 November 1871, the length of the barrel and the stock fittings had been approved, but the absence of a safety mechanism was criticized. On 9 December 1871, however, the single-shot bolt-action Mauser was provisionally adopted.

The Prussian government arsenal in Spandau was instructed to make 2,500 'Interims-Modell' or pre-production rifles to facilitate field trials, and two alternative safety-catch designs were shown to the *Gewehr-Prüfungs-Kommission* on 14 February 1872. Selection of a winged spindle on top of the cocking-piece, which rapidly became a Mauser trademark, allowed Kaiser Wilhelm I to sign the decree adopting the 'Infanterie-Gewehr Modell 1871' as the standard infantry weapon of the German Empire on 22 March 1872.

Though the bolt mechanism was the work of the Mauser brothers, the calibre, barrel and the design of the rifling had simply been copied from the Chassepot. The trigger was an adaptation of the 1862-pattern Dreyse infantry-rifle mechanism, whilst the stock and fittings had been devised by the *Gewehr-Prüfungs-Kommission*. The perfection of the 11mm centre-fire cartridge, the extractor and the ejector mechanism was credited to a governmental committee chaired by *Oberst* von Kalinowski.

THE 1871-PATTERN MAUSER RIFLE

The introduction of the Mauser rifle to general service was delayed by problems. The formation of the German Empire in 1871, with its threat of Prussian domination, had traumatized many of the smaller German states; the removal of border controls and customs tariffs, in conjunction with the opening of markets to free trade, soon damaged the economy of the fledgling empire.

A serious financial crisis nearly brought a catastrophe that was averted only by a series of unpopular measures. In addition, excepting parts of Saxony, none of the German states – Prussia included – were industrialized. Manufacturing capacity lagged behind Britain, France and even Austria-Hungary.

Apart from the Prussian government factory in Spandau and the Bavarian establishment in Amberg, only the Dreyse factory in Sömmerda could make breech-loading rifles in quantity. The Mauser was a sophisticated design by the standards of the day, and the need to make guns in hundreds of thousands – with parts which could be interchanged at random – was beyond the capabilities of most German engineering companies. Shortages of machine-tools were a particular weakness which could be solved only by placing orders with companies such as Pratt & Whitney in the USA and, to a lesser extent, Greenwood & Batley in Britain.

Production of 1871-type Mauser rifles was supposed to have started in the Danzig, Erfurt and Spandau factories in 1872–73, but work was delayed by the shortages of machine-tools and an assortment of teething troubles; the Spandau plant, for example, did not begin work until the spring of 1874. Consequently, although the first series-made M1871 rifle was exhibited before Kaiser Wilhelm I on 22 March 1875 to mark the third anniversary of its adoption, the Prussian *Kriegsministerium* realized that re-equipping the Prussian Army within the five-year target could be achieved only by recruiting additional contractors.

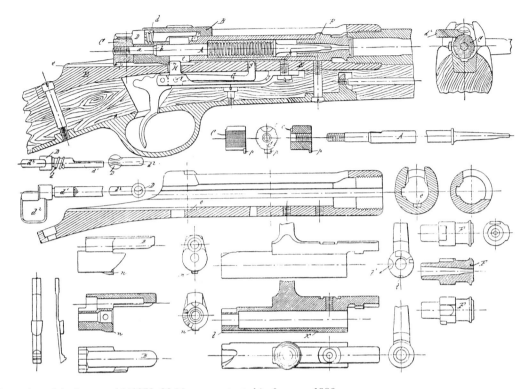

The action of the improved M1878–80 Mauser, patented in January 1880.

A hundred thousand rifles had been ordered from the Bavarian factory in Amberg in May 1872, but converting Werder rifles to chamber the 11mm *Reichspatrone* dragged on until the mid-1870s and the first Bavarian Mausers were not made until 1877.

A Prussian contract for 500,000 guns was given to Österreichische Waffenfabrik-Gesellschaft ('Waffenfabrik Steyr') in the summer of 1873, whereas Württemberg ordered 100,000 rifles from Gebrüder Mauser in December 1873. Ironically, the Mauser brothers were allowed to purchase the Oberndorf rifle factory from which Paul (as he was widely known) had been sacked seven years earlier. Gebrüder Mauser & Cie delivered the last gun of the initial order in July 1879.

A Prussian contract for 180,000 guns went to a cartel of Spangenberg & Sauer, Schilling and Haenel in February 1876; another for 100,000 was given to Österreichische Waffenfabriks-Gesellschaft in mid-1876; and a third, for 75,000, went to the National Arms & Ammunition Co., Ltd in Birmingham, England. Only about 6,000 British-made Mauser rifles were ever completed. By comparison, Österreichische Waffenfabrik-Gesellschaft delivered 474,622 rifles, 60,000 carbines, 150,000 bolts, 55,963 receivers and 52,000 barrels to Prussia and Saxony prior to 31 December 1877, and it is clear that total production of 1871-type rifles must have exceeded one million. Mauser sent 26,000 rifles and carbines to China in 1876, while Serbia placed an order for 100,000 similar M78/80 or 'Mauser-Koka' rifles in 1878.

TRIALS AND TRIBULATIONS

The performance of the Mauser was a revelation to troops accustomed to obsolescent Dreyse needle guns, once the high point of military breech-loaders but long past their prime by 1875. The 11mm *Reichspatrone*, with a 24.6g round-nose bullet driven by a charge of nearly five grams of black powder, gave a muzzle velocity of 440m/sec. According to British experiments, the projectile was still moving at 859ft/sec at 500yd, dropping to 629ft/sec at 1,000yd, 459ft/sec at 1,500yd, and 338ft/sec at 2,000yd. This compared reasonably well with rival European designs.

Service use soon showed that the Mauser was beset with problems. Cartridges failed to ignite satisfactorily, accuracy was poorer than anticipated, and the bolt-stop washer screw worked loose. The certainty of ignition was considerably improved by altering the striker spring, but the rifles were always dogged by accuracy problems ascribed to asymmetry in the locking system.

The *Infanterie-Gewehre* M1871 served the regular forces until the first large-scale deliveries of 1871/84-type magazine rifles were made in the mid-1880s. After serving the pioneers, the Train and associated second-rank units, the single-shot Mausers were handed from the regular army to the *Landwehr*, and then from the *Landwehr* to the *Landsturm* by the 1890s.

Many were sold out of service as better weapons became available, but surprising numbers reappeared during World War One. A few survivors are said to have had their barrels bored-out to 13mm and been issued with incendiary ammunition as anti-balloon guns, but this claim remains controversial. Surplus 1871-type Mausers sold briskly into the South American and Far Eastern export markets late in the nineteenth century, and a modified version, altered in France to chamber the 6.5 × 53.5mm No. 12 Daudetau cartridge, was used in Uruguay for some years.

THE FIRST MAGAZINE RIFLES

Though the advantages of magazine feed had been conclusively demonstrated by Winchester-armed Turks during the Battle of Plevna (1877), conser-

vatism was as much a part of the late-nineteenth century German Army as any other military power.

By 1880, however, it was clear even to its most hidebound devotee that the single-shot Mauser rifle had been overtaken by better designs. Most importantly, the French Navy had purchased a substantial quantity of Austrian-made Kropatschek magazine rifles in 1878–80, and the French Army was known to be experimenting with adaptations of the same basic design. Rivalry between Germany and France had assumed epic proportions by this time, and a step forward in one army was customarily followed by a riposte from the other.

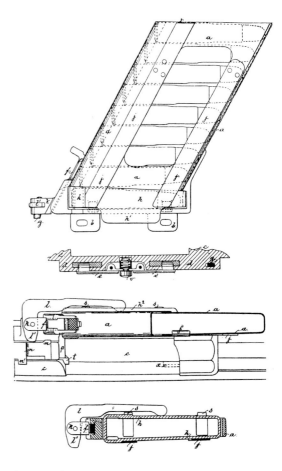

A gravity-feed box magazine for the Mauser rifle, patented in May 1887. R.H. Korn

Several experimental magazine conversions had been developed for the single-shot Mauser in the late 1870s and early 1880s, but none attained service. Korn pictures two in *Mauser-Gewehre und Mauser-Patente* (1908), one with a curious side-mounted box magazine and another with a saddle running down and under the stock. Among other attempts to develop auxiliary magazines for the Mauser were those of Carl Holub, inventor of the Austrian Werndl rifle, whose tubular design dated from 1878; Franz von Dreyse, son of the inventor of the needle-rifle, patentee of a tube magazine in 1879 and a more conventional box pattern in 1882; Louis Schmeisser, who developed a drum unit in 1881–82; Ludwig Loewe & Co., whose saddle unit dated from the early 1880s; and Edward Lindner, who contributed a gravity-feed box in 1883.

Development of a magazine-fed rifle began in Oberndorf early in 1881 in response to government requests. Patented in March 1881, the prototype Mauser-*Probegewehr* C81 was adapted from the Serbian 1878/80-model infantry rifle. It had a new

receiver, much deeper than the 1871 pattern, which contained a bolt-operated tipping cartridge elevator.

The tube magazine beneath the barrel was scarcely innovative; the Henry rifle had appeared in the USA at the beginning of the American Civil War, and the 10.4mm rim-fire Vetterli rifles had been serving the Swiss Army since 1869. Even the French Navy had been using the Austrian-designed Kropatschek since 1878.

The 1881 patent drawings illustrate several different elevator designs, but none of the prototypes was successful and a patent of addition was filed in May 1882 to protect refinements to the breech mechanism. Trials proceeded quickly, amid worries that the French were stealing a technological lead, and 2,000 examples of the experimental *Infanterie-Repetier-Gewehr* C82 made by Gebr. Mauser & Cie were issued in the summer of 1882 to infantry battalions garrisoned in Darmstadt, Königsberg in Preussen and Spandau.

Though the trials were successful enough, a few cartridges had exploded in the magazines. This was due to the rounded nose of the standard 11mm

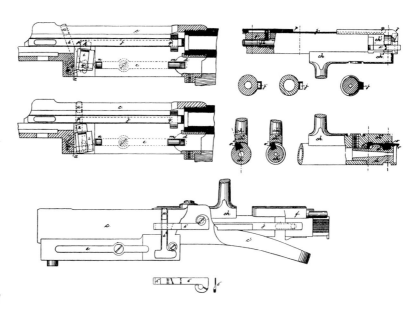

The bolt and receiver of the C82 rifle, from a patent granted in May 1882. R.H. Korn

Paul Mauser (1838–1914). Mauser-Werke Oberndorf GmbH

Wilhelm Mauser (1834–88). Mauser-Werke Oberndorf GmbH

bullet slamming into the primer of the round lying ahead of it in the tube; potentially very dangerous, the problem was eliminated by introducing flat-nosed bullets and recessing the primers.

The 11mm *Reichspatrone* was retained for the new rifle, saving time and money, although trials undertaken in April 1879 and the autumn of 1883 had demonstrated the advantages of 9mm cartridges – the smallest calibre acceptable for use with black-powder propellant. Ammunition problems excepted, the worst feature of the experimental magazine rifle was the inevitable change in the point of balance as the cartridges in the magazine were expended.

The new rifles passed their adoption trials satis-factorily, and, once minor amendments had been made to the extractor and the ejector, became the *Infanterie-Gewehr* M1871/84. 'His Majesty the King-Emperor,' stated a *Kriegsministerium*

memorandum of 22 January 1884, '… is pleased to order that production of the M1871 infantry rifle is to be suspended… It is [his] decision that, in the interests of secrecy, the designation of the new pro-totype will be M71/84 and that the expression "Repeater" will not be applied …'

Paul Mauser was keen to safeguard his rights, and a licensing agreement was concluded on 22 July 1885 between the Prussian government (with power of attorney for the states of the German Empire), Paul Mauser and Waffenfabrik Mauser AG. A three-Mark royalty was to be paid on each of the first 100,000 guns to pass proof, and one Mark on each rifle made thereafter. In return for a guaranteed sum of 300,000 Marks, Mauser agreed to forego royalties on guns made before the licence had been signed and, if appropriate, to correct design faults without charge.

The Mauser 'Oberes Werk' (Upper Factory) in Oberndorf, drawn in 1909. Dr Rolf Gminder

The first issues were made in 1886 to XV and XVI Corps of the German Army, guarding the borders in Alsace-Lorraine. No sooner had the Mauser reached the armed forces than the French announced the introduction of the small-calibre Mle 1886 Lebel rifle. Panic spread in German ordnance circles. The new rifle was similar to the *Gewehr* M1871/84 – little more, indeed, than a Kropatschek – but the introduction of a cartridge loaded with smokeless propellant was a great leap forward.

Experiments with propellant derived from nitro-glycerine had been underway in most European countries for more than twenty years, and the French were the first to perfect a manufacturing process suited to military ammunition. For once, France had caught Germany unprepared.

In March 1887, however, a French deserter – for a price – offered the Germans a Lebel rifle and enough 8mm ammunition to enable the constituents of the propellant to be analysed. The search had begun for a brand-new magazine rifle to tip the balance of power back from France to Germany.

THE FIRST SMALL-BORE RIFLE

In November 1887, the *Gewehr-Prüfungs-Kommission* contacted selected officers and technicians attached to the arms factories in Danzig, Erfurt and Spandau, and asked them to consider two alternatives: converting existing stocks of M1871/84

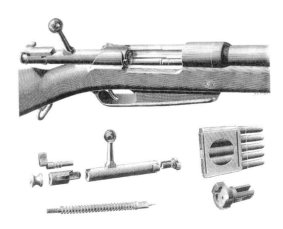

The Gewehr 88. Engineering

37

The action of the Belgian Mle 1889 Mauser. Ian Hogg

rifles for an 8mm cartridge loaded with smokeless powder being developed by Rottweiler Pulverfabriken, or developing an entirely new weapon. To save even more money, the existing tubular magazine could be retained if the new cartridge was to be little more than a necked-down version of the old 11mm *Reichspatrone*; only the sights would have to be revised.

The consensus of opinion suggested that the *Gewehr* 71/84 should be altered by the addition of a *Mauser-Verschluss mit doppeltem Widerstand* (Mauser bolt system with double locking lugs), and production machinery was ordered from Ludwig Loewe & Cie in December 1887.

Leading German small arms experts soon became sceptical of the potential in the modified *Gewehr* 71/84. Disquieting rumours were appearing in the press, and advances being made elsewhere were being mentioned even outside military circles. Worried, too, about the penny-pinching attitude of the government, the German specialists suggested that trials should be undertaken with more efficient weapons.

The *Gewehr-Prüfungs-Kommission* began by revising the lock-work and bolt mechanism of the M1871/84 rifle, much of the work being credited to Louis Schlegelmilch of the Spandau arsenal. The prototype *Reichsgewehr* or 'commission rifle' was a

Prussian artillerymen pose with a captured Belgian howitzer at the Brasschaet shooting range, 10 January 1915. Note the piled Gewehre 88 *to the right.*

strange hybrid, with a Schlegelmilch-Mauser action, a clip-loaded magazine inspired by Mannlicher, and a barrel jacket designed by Armand Mieg. Copying the pitch and rifling profile of the French Mle 86 rifle soon proved to be a mistake, but it undoubtedly minimized development time.

The first experiments with the new German rifle were undertaken early in 1887 with rimmed 8mm cartridges adapted from the 11mm *Reichspatrone*. By midsummer, however, a modified Swiss-type rimless design had been issued for trials and, on 23 March 1888, the Bavarian military observer in Berlin, General von Xylander, was able to report that the rifle design had been virtually completed. The prototype was immeasurably superior to the obsolescent *Gewehr* 71/84 and its twin-lug derivative; it was also a better combat weapon than the Lebel.

THE *JUDENFLINTEN-AFFÄRE*

The Germans do not seem to have appreciated that the clip-loaded magazine of the *Gewehr* 88 was clearly based on Mannlicher's design, nor had arrangements been made to compensate Armand Mieg for the barrel jacket which had been patented with his experimental small-bore rifle in 1887.

On 20 March 1889, therefore, the Prussian government granted Mieg 50,000 Marks in recognition of his contribution to small arms design. Yet disputes continued, and an argument over the infringement of Mannlicher patents was resolved only by permitting Österreichische Waffenfabriks-Gesellschaft to offer 'Mannlicher' rifles commercially on the basis of *Reichsgewehr* actions.

Trials with about twenty 'pre-production' *Reichsgewehre*, hand-made in Spandau in the spring of 1888, continued until the calibre question had been resolved in favour of the 8mm cartridge. Work was completed by July 1888, mass-production plans were prepared, and limited series production began in October of the same year.

Encouraging field-trial reports persuaded the *Gewehr-Prüfungs-Kommission* to recommend immediate adoption of the experimental rifle.

Orders were duly signed by Kaiser Wilhelm II on 12 November 1888, production began precipitately, and the first issues were made to the army corps in Elsass-Lothringen in the spring of 1889.

Rumours concerning breech explosions began to circulate almost as soon as the guns had entered service, and many commentators asked why Paul Mauser, Germany's premier rifle designer, had been spurned by the army authorities. The story took an ugly turn when Hermann Ahlwardt, a Berlin schoolmaster/pamphleteer, tried to blame the shortcomings on Ludwig Loewe & Cie.

Loewe, Ahlwardt claimed, had bribed government arms inspectors to accept inferior workmanship and was profiteering from deliberate attempts to undermine the German Army. Differing accounting procedures should have prevented comparisons being made between the output of private companies and the state-owned arms factories.

It was no coincidence that the Loewe family and Max Duttenhofer, owner of the Rottweiler propellant-making business, were Jewish: the underlying motive of the accusations was rabid anti-Semitism. Whether the rumours had originated in the business community or right-wing factions within the *Reichskriegsministerium* was less obvious. But the press gleefully exploited the story, and the *Reichsgewehr* gained the sobriquet of *Judenflinte* ('Jewish Rifle').

The truth was more prosaic. The *Gewehr* 88 was not a particularly bad design; indeed, it was better in many respects than the Lebel, the 1888-type Mannlicher and the Lee-Metford. There were teething troubles (explained in detail in Part Two), but most of them were speedily overcome. The rifle had simply been introduced into service too quickly and lack of experience with the chemistry of smokeless propellant proved to be a great handicap.

The necks and bases of the earliest cartridge cases were prone to split in winter cold or summer heat (a phenomenon known as 'season cracking'), but adding stab crimps to retain the bullet and the primer minimized the problems. Annealing the necks eventually eliminated cracking altogether, simultaneously improving the seal between the case and the bullet.

Changes were made to the thickness of the neck and walls of the cartridge case, and the flash-holes were drilled instead of punched. A nitro-cellulose flake propellant developed in the Spandau ammunition factory in the mid-1890s was introduced to prevent solvents in the propellant evaporating during storage and raising chamber pressures to dangerous levels. However, the last problems were only solved by the approval of an *Einheitshülse* (universal case) in 1901. Intended for use in rifles and machine-guns interchangeably, cartridges of this type included an additional 'E' in their headstamps.

The charge that the Germans ignored Paul Mauser to their cost is also worth scrutiny. A small-calibre bolt-action prototype had been patented in Germany in April 1888, but handling a similar rifle supplied to British trials shows that it had notable weaknesses. Lugs locking in the receiver-bridge placed the front part of the bolt under great stress when the gun was fired; the cocking-piece was so cumbersome that a normal hand could not grasp the stock and trigger comfortably; and the design of the magazine was poor.

Although this curious weapon metamorphosed into the efficient Belgian Mauser service rifle, the most important changes were not made until after series production of the *Reichsgewehr* had begun.

SMALL-BORE MAUSERS

Experiments began in 1892 with bullet diameters as small as 5mm, seated in cases measuring 67–68mm developed by Polte of Magdeburg. The cartridges were fired from modified *Reichsgewehre*, but the trials had few lasting results other than to demonstrate shortcomings in the rifle design.

By this time, the clumsy C88 Mauser had been superseded by the Belgian Mle 1889. The first of the charger-loaded small-bore magazine Mausers, the Mle 89 had a distinctive single-column magazine protruding beneath the stock ahead of the trigger.

Improvements were soon made in the Mauser rifles. A fixed non-rotating extractor was introduced in 1890, followed in 18922 by a striker held in the bolt plug by lugs. A staggered-row magazine carried entirely within the stock appeared in 1893; a three-position safety lever was added on top of the bolt plug in 1894; and a distinctively flanged bolt plug with a bolt-sleeve lock was introduced in 1895. A special method of breeching was also patented in 1895, interposing a collar machined integrally with the receiver between the bolt head and the barrel-face.

Development of the 'Spanish' or 1893-model Mauser, with its internal staggered-row box magazine and a reputation for unerring accuracy, proved a turning-point. These guns were to prove themselves in the Spanish-American War of 1898, countless South American revolutions, and the Second South African ('Boer') War of 1899–1902 – inspiring the old British superstition that lighting three cigarettes from a single match was unlucky.

STEPS TO PERFECTION

Experiments began in November 1894 when 200 rifles with new Mauser actions and charger-loaded magazines were issued for troop trials. The special tangent backsight credited to *Oberstleutnant* Lange, a one-time director of the Spandau munitions factory, became a characteristic of pre-1918 German service rifles.

The *Gewehr-Prüfungs-Kommission* subsequently ordered 2,000 of its improved rifles from Waffenfabrik Mauser AG in January 1895 and a letter written by a Bavarian observer confirms that a specimen gun had been prepared by 5 February 1895. A thousand of each of two models was delivered in the early summer of 1895, the two types differing in breeching arrangements. They were issued on 1 August 1895 to the fusilier battalion of 1. *Garde-Regiment zu Fuss*, the *Garde-Jäger-Bataillon*, the third battalion of *Füsilier-Regiment* Nr 86, and the *Militär-Schiess-Schule* in Spandau.

The Prussian government acquired the rights to exploit patents protecting the *Mauser-Schloss und Magazineinrichtung* (Mauser lock and magazine design) on 16 November 1895. The licence was to run for seven years, paying a one Mark royalty on each of the first 100,000 guns and 50 pfennige (half a Mark) thereafter.

Trials showed that the new Mauser was much more efficient than the *Reichsgewehr*. Its 1895-patent action also proved superior to the 1893 model incorporated in the experimental rifles purchased from Mauser in 1896 (qv). Finally, the Kaiser approved adoption on 11 March 1897 as the *Gewehr 88/97* and the king of Bavaria signed the appropriate papers on 21 April 1897.

Two thousand *Gewehre 88/97* were made in Mauser's Oberndorf factory prior to July 1895, embodying features which were new to the German Army. The most important patent was granted on 30 October 1895 to protect a third locking lug and a cocking-piece housing with integral gas deflector lugs. Gas escape holes were bored in the bolt body, and gas could also vent through the charger-loading cutaway milled in the left side of the receiver by way of the left-hand locking lug guide-way.

The rifle had a charger-loaded staggered-column box magazine, contained within the stock, and measured about 1,240mm overall; weight averaged 4kg. Its 740mm barrel was rifled with four grooves twisting to the right, and the Lange-pattern backsight was graduated from 300m to 2,000m.

Although the action greatly resembled that of its 1898-type successor, with a straight bolt-handle and a cutaway milled in the left side of the receiver wall, the M1888/97 rifle had a prototype *Lange Visier* mounted on top of the barrel jacket. It lacked a hand guard, had a straight-wrist butt, and the extraordinary double-clamp nose-cap and bayonet bar assembly was based on a patent granted to Mauser late in 1895.

A half-length cleaning rod protruded from the centre of the cylindrical bayonet-attachment bar, and the true muzzle protruded a short distance from the barrel jacket. A solitary screw-clamping spring-retained barrel band carried the front sling swivel; a second swivel being mounted either through the front of the trigger guard bow or beneath the butt.

The Erfurt small-arms factory was asked to attain a daily production rate of 130 guns as soon as possible, but field trials indicated that the *Gewehr 88/97* fell short of perfection and its introduction into service was cancelled. Its replacement, the *Gewehr 98*, discarded the barrel jacket and the peculiar bayonet attachment assembly, but gained a pistol-grip stock to help the firer control recoil and had an improved backsight.

Even though trials were underway, the *Gewehr-Prüfungs-Kommission*, mindful of moves being made elsewhere, ordered 2,185 experimental 1896-pattern Mauser rifles. Most of these chambered a 6mm cartridge, as the German Army was keen to assess the potential of small-calibre projectiles.

Ordered in 1896 and delivered in 1897, the Mausers lacked barrel jackets but used an older action than the *Gewehr 88/97*. Tangent-leaf backsights were used instead of Lange tangent patterns, and the bayonets were attached to the muzzle conventionally. However, so many rifles were re-chambered or tested to destruction that their development history is obscure.

Rifle No. 53 chambered a 6mm round, was 1,250mm long, weighed 3.63kg empty, and had a 740mm barrel rifled with six grooves making a turn to the right in 165mm. Rifle Nos 74 and 250 may have been similar, though Korn – listing their shooting results in *Mauser-Gewehre und Mauser-Patente* – describes them as '*Gewehre 98*'; Rifle No. 9 seems to have fired a 7.65 × 53mm cartridge.

Mauser and Deutsche Metallpatronenfabrik developed the special 6mm ammunition for the *Gewehr-Prüfungs-Kommission*. Published dimensions suggest that it may have been based on the US M1895 (Lee Straight Pull) pattern, DM/DWM case number 425A; the lists also state that case 425B was an experimental 6mm-calibre German cartridge.

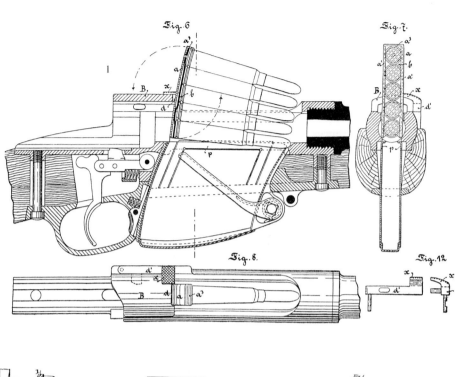

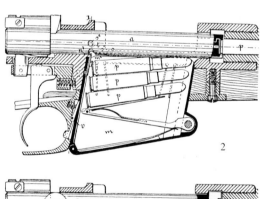

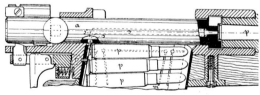

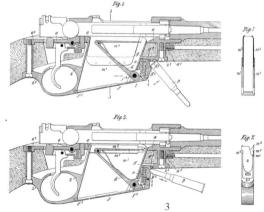

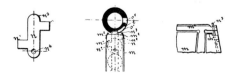

Three stages to the Gewehr 98: *magazine patents granted in October 1888 (1), November 1890 (2) and June 1892 (3).* R.H. Korn

The Mauser rifles delivered to the *Gewehr-Prü-fungs-Kommission* in 1896–97 were similar to the 1896-model Swedish guns, but had been modified in accordance with patents granted in the early 1890s. The most important was probably DRP 65,225 of February 1892, which protected the non-rotating extractor. The action was locked by two vertical lugs on the front of the bolt body, and a third lug (under the bolt body) which engaged a recess milled in the receiver behind the magazine well.

A slot in the left locking lug allowed the ejector blade to kick the extracted cartridge out of the feed way. The bolt head was partly shrouded, gas-escape ports were bored in the bolt body, and a spring collar anchored the non-rotating extractor. The safety lever – a Mauser 'wing' design – allowed the rifle to be fired if the wing was rotated to the left, though the striker was locked when the blade was vertical, and the striker and the bolt were immobilized with the wing rotated towards the right.

The internal box magazine contained a staggered row of five cartridges, and had a detachable floor plate. Guides milled in the front portion of the solid receiver-bridge accepted the C94 charger, and a large thumb-clearance cutaway was milled out of the left receiver wall. The trigger could not release the sear until the bolt was securely locked, when a small lug on the sear bar entered a small recess in the underside of the bolt body; only then could the sear nose be rotated to release the striker.

The one-piece walnut stock had a straight wrist instead of a pistol-grip, although some rifles were subsequently re-stocked in the manner of the *Gewehr* 98 (qv). The bolt-handles had spherical grasping knobs and were generally turned down against the stock. The barrels lacked jackets, but walnut hand guards ran forward from the receiver ring to protrude ahead of the solitary screw-clamping barrel band. The band was held by a spring and carried a sling swivel. The rear swivel lay beneath the butt.

The nose-cap was a small clamping band with a bayonet lug on the right side; a half-length cleaning rod, for emergency cleaning and removing obstructions from the bore, protruded from the fore-end under the muzzle; and a tangent-leaf backsight was mounted on a short sleeve around the barrel. Metal parts, excepting the grey-pickled receivers, bolts and butt plates, were blued and polished. Most guns had the Mauser name above the chamber, and commercial proof marks on the barrel and receiver.

The *Gewehr-Prüfungs-Kommission* trials lasted from 1895 until about 1902, but soon showed that the 1895-patent Mauser action was an improvement on the altered 1893 type embodied in the '1896' rifles. The *Infanterie-Gewehr Modell* 1898 (*Gewehr* 98 or 'Gew. 98') was adopted in April 1898, but testing of different cartridges, rifling profiles and twist-pitches continued for several years. Calibres of 6mm, 6.5mm, 7mm, 7.65mm and 8mm were among the many to be tried, but the flat trajectory and high velocity of the 6mm bullets could not overcome their inferior lethality and wind-riding qualities.

THE *GEWEHR* 98

The perfected 1898-type Mauser was among the most reliable rifles issued for service prior to 1918. Cutting away enough of the bolt head rim to allow cartridge rims to rise under the extractor as they left the magazine effectively eliminated double-loading. Arrangements to deflect gas from a pierced primer away from the firer were very good, and safety arrangements were praiseworthy. Extraction was greatly assisted by the camming action of the helical entrance to the locking-lug seats; and machining the feed lips into the underside of the receiver gave a reliable feed. The Mauser was also cocked on opening the bolt – unlike many rivals – and could even be re-cocked in the event of a misfire merely by raising and then lowering the handle.

The true 1898-type action included two locking lugs on the head of a solid-body bolt, bored out from the rear; a third or 'safety' lug under the rear of the bolt, ahead of the bolt-handle base; and a non-rotating extractor attached to the bolt body with a broad collar.

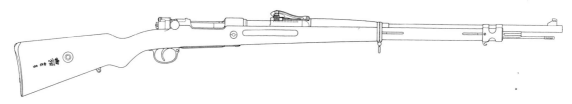

The Gewehr 98. John Walter

A modified firing pin protected by DRGM 154,915 of 22 May 1901, introduced on military rifles in 1902, added flat-faced lugs on the head of the striker. These lugs could enter recesses inside the bolt head only if the bolt-handle was at least partly closed, but the purpose of this feature has been questioned and may simply have prevented ignition if the firing pin shaft broke or the cocking-piece detached while the bolt was being closed.

The trigger consisted of just five parts: the trigger lever, a rocking sear pivoted at the front (beneath the receiver), a small coil-spring and two pins. However, though ideal militarily, the trigger was usually the first item to be discarded in sporting rifles.

The left side of the receiver was cut away in front of the receiver-bridge in accordance with DRGM 56,068 of 9 August 1895, to allow the thumb to press cartridges from the charger into the magazine. However, cracks may appear in this area on 1898-type actions made during World War One. Slow lock time was another poor feature.

About 90,000 guns had been received from all the contractors involved by late 1901. The first issues were made to the *Kaiserliche Marine* (Imperial Navy) and the *Ostasiatisches Expeditionskorps*, and then to the first three Prussian Army corps in the autumn of 1901; the next three corps re-equipped in 1902, three more in 1903, and the process continued until issue was complete.

Budget provisions for the *Kaiserliche Marine* in 1905 suggested that 50,873 rifles – 6,803 *Jägerbüchsen* 71, 17,145 Gew. 71/84 and 26,925 Gew. 98 – would satisfy the needs of the active land-based units, with an additional 15,000 guns aboard the warships.

By 1907, many of the obsolescent *Jägerbüchsen* 71 had been exchanged for *Gewehre* 98 and more than 32,000 modern Mausers had been distributed to land-based units, ranging from only 289 in III. *Matrosen-Artillerie-Abteilung* to 10,544 in I. *Matrosen-Division*. The inventory also contained 16,639 Gew. 71/84 and 4,985 *Jägerbüchsen* 71, though the latter rifles had been withdrawn from active service by 1908.

Many *Gewehre* 98 were sent to the battle fleet in exchange for older weapons. Total naval land-service issue of *Gewehre* 98 had declined to 18,349, with only the *Seebataillone* retaining their 1907 issue levels; pistols had even replaced rifles entirely in the *Torpedo-Divisionen*.

The budget for the 1910 financial year noted the requirements of the land-based naval units as 26,968 Gew. 71/84 and 23,759 Gew. 98. Scales of issue laid down in the Tirpitz Principles (1909) suggest that at least 20,000 guns would have been carried aboard warships, so the *Kaiserliche Marine* would have had about 50,000 modern Mausers by the beginning of World War One. The German Army probably had twenty times as many.

THE AUTOMATIC PISTOL

By 1890, the military authorities had realized that advances in small arms design had overtaken the robust (but primitive) commission-designed service revolvers of 1879 and 1883. Tests were being undertaken by the *Gewehr-Prüfungs-Kommission* as early as 1891, and procurement of revolvers had even been suspended while the merits of the new 'repeating pistols' were assessed.

Early trialists included the Borchardt, the Mieg, the Mauser and the '*Spandauer Selbstladepistole M1896*', though a few lightened double-action revolvers were made experimentally in the Erfurt rifle factory in 1896–97 in case the self-loading concept was abandoned altogether.

The first auto-loading pistol to interest the German authorities was the Mauser *Selbstladepistole C96*, designed by the three Feederle brothers in 1894–95 but now customarily credited to Paul Mauser himself. This mistaken attribution has arisen from the granting of the patents in the name of Paul Mauser alone; typical of the days when patents were sought corporately and designers were all too often kept in anonymity.

The Mauser pistol was operated by allowing the barrel and the receiver to run back until, when the chamber pressure dropped to a safe level, a locking block pivoted downward to release the bolt. Movement of the barrel and receiver stopped to let the bolt reciprocate alone, cocking the hammer and stripping a new round into the magazine as it returned. The lock pivoted upward, and then the whole assembly ran back to battery.

The pistol was powerful, but also clumsy and badly balanced. Its cartridge was simply a controversial adaptation of the Borchardt pattern of 1893. Waffenfabrik Mauser was owned at that time by Ludwig Loewe & Cie, the original promoters of the Borchardt pistol, and it is suspected that Mauser's social standing allowed him to appropriate the Borchardt cartridge simply by gaining Loewe's permission. Hugo Borchardt reportedly saw this as unduly favouring a competitor at the expense of his own gun.

Whatever its drawbacks may have been, the Mauser C96 was an engineering marvel. The pieces interlocked without screws and, as might be expected of Waffenfabrik Mauser, no expense was spared in its construction.

Trials revealed a tendency for cartridges to jam in the breech, though tests undertaken in 1896 by the *Gewehr-Prüfungs-Kommission* had been successful enough to encourage the purchase of 145 guns for field trials. Delivered in June 1898, these were subsequently issued to the *Infanterie-Schiess-Schule* (School of Marksmanship), the *Garde-Jäger zu Pferde* and the *Leib-Garde-Husaren-Regiment*.

Shortly before Christmas 1898, the *Gewehr-Prüfungs-Kommission* told the *Kriegsministerium* that the Mauser C96 pistol was greatly preferable to a revolver. It shot more accurately, and the high-velocity bullet had a flat trajectory. Immediate adoption could not be recommended, owing to mis-feeds, jams, and the unsuitability of the stock and holstering arrangements, but 124 additional pistols were ordered in January 1899. These were

A typical Mauser C96 pistol, with a fixed standing-block backsight and a six-round magazine in the frame ahead of the trigger-guard. Joseph Schroeder

issued to two infantry regiments (Nos 48 and 72), two *Uhlanen* (lancer regiments; Nos 5 and 13), and the *Fussartillerie-Regiment* Nr 3.

The state army of Württemberg acquired forty-eight Mauser C96 pistols of its own, but Saxony and Bavaria decided simply to monitor developments in Prussia. Reports of the trials reached the *Kriegsministerium* in November 1900, but repeated the verdict of the 1899 series: the gun was powerful, but jammed too frequently to justify being adopted.

In February 1901, realizing that the trials with the Mauser C96 were still failing to meet expectations, the *Kriegsministerium* ordered tests of a Mannlicher and the 1900-pattern Borchardt-Luger to begin. Deutsche Waffen- und Munitionsfabriken supplied a small number of 120mm-barrel Swiss-type 7.65mm Borchardt-Lugers shortly afterwards, but the *Gewehr-Prüfungs-Kommission* remained keen to persist with the C96 until the reputation for jamming became too strong.

By midsummer 1901, the *Kriegsministerium* noted that the Mauser and the Borchardt-Luger were still being considered, though the Mannlicher had been dismissed as unsuitable. As the C96 was still suffering feed problems, Mauser was asked to supply two modified guns, one with a short barrel and the other a 'reduced-weight type for officers'; greater reliability was essential.

These improved Mausers were far superior to their predecessors; one had even fired two series of 500 rounds flawlessly, with neither oiling nor cleaning. Although the German authorities had virtually decided to adopt the Luger by 1904, the C96 – very popular commercially – remained in production until the beginning of World War Two; the 100,000th example was sold in 1910.

THE FIRST MACHINE-GUNS

The German Army had never been particularly enthusiastic about the mechanical machine-gun, owing to the poor performance of the Bavarian 'Feld' machine-gun and the French *Mitrailleuse* in the Franco-Prussian War. A few Gatlings were purchased for experimental purposes in the 1870s and 1880s, but little else was done.

In 1887, however, during a visit to Britain to celebrate the Golden Jubilee of his grandmother, Queen Victoria, Kronprinz Wilhelm – soon to become Kaiser Wilhelm II – visited the 10th Hussars with a group of cavalry officers. He was so impressed by a multi-barrel Nordenfelt machine-gun, mounted on a two-horse 'galloper' carriage, that a gun of identical pattern was immediately ordered for trials in Germany. The colonel of the 10th Hussars charitably sent an experienced gunner over to Potsdam to train German cavalrymen.

When the Maxim appeared in the mid-1880s, enthusiasm for machine-guns was still muted even

An artist's impression of the 'galloper' Nordenfelt gun in use with the German cavalry.

though a British-made gun was successfully demonstrated in Spandau in 1889. In 1895, however, Ludwig Loewe & Cie imported a gun from the Maxim-Nordenfelt Guns & Ammunition Co., Ltd, converted it to fire the standard 8mm 1888-pattern service cartridge, and submitted it for trial. Duly impressed, the *Gewehr-Prüfungs-Kommission* recommended the purchase of additional guns for field trials.

Loewe had meanwhile acquired a manufacturing licence which permitted Maxim-type guns to be sold throughout much of Europe, excepting Greece, Portugal and Spain, where rights had already been granted. Lapse or purchase had annulled a previous agreement between Maxim-Nordenfelt and Krupp. The formation of Deutsche Waffen- und Munitionsfabriken at the beginning of 1897 encouraged the production of Maxims to continue in the former Loewe factory in Hollmannstrasse, Berlin.

The first German guns, incorporating many British-made parts, were supplied to the *Kaiserliche Marine* in 1897–98. Mounted on simple British-style tripods, they had water-jackets, fusee-spring casings and back plate/handgrip assemblies made of corrosion-resistant bronze. Maxims were particularly favoured for warships stationed overseas, and by the *Kreuzergeschwader* (Cruiser Squadron) formed in 1898 to protect German colonies in the Pacific. Standard issue was four machine-guns per cruiser-size ship.

Although the German Army had tested Maxims experimentally in 1896–97, little more was done until four-gun batteries attached to the *Jäger-Bataillon* serving the cavalry division were tested at the *Kaisermanöver* (Imperial manoeuvres) in 1899. At much the same time, Kaiser Wilhelm II bought a Maxim for each of the *Garde-Dragoner-Regimenter* (guard-dragoon regiments) on his own account.

Machine-guns were attached to cavalry units for extended field trials in 1900; in 1901, they were successfully tried by cyclists' detachments attached to the cavalry divisions. The success of these experiments led to the creation of the first five machine-gun detachments, one serving the *Gardekorps* and the others being attached to individual army corps headquarters. Finally, in 1902, a machine-gun battery was attached to each cavalry brigade in frontier districts and eight batteries were formed to serve army corps stationed in the interior.

Batteries customarily consisted of four Maxims, each worked by four-man teams. Fifteen thousand rounds of ammunition were carried on the limbers and supply carts for each gun.

The original guns are now customarily designated *Maschinengewehr* 99, but this term came into use only after the introduction of an improved pattern in 1901. Army-issue MG01 Maxims were much the same as their *Kaiserliche Marine* predecessors, but the transverse hole for the tripod-cradle retaining pin (in the lower front of the receiver) was

A Maxim Gun on a wheeled carriage, typical of the examples tested in many countries prior to 1900.

The MG01 could be dragged for short distances on its sled mount.

replaced with lugs on the rear of the brass water jacket which engaged a gimbal on the mount.

Though tripods and pintle mountings were tried, the perfected design took the form of a wheeled sled designated *Schlitten* 01. A special quadrant elevator was fitted to the sled body, but the traversing arrangements were very limited. The design of the sled allowed the gun to be dragged short distances over open ground, once the legs had been folded out of the way.

Trials revealed weaknesses in the design, and the *Schlitten* 01 was replaced by the *Schlitten* 03. This had an improved elevation system, substantially greater in-built traverse, and a more robust connection between the gun and the mount cradle. The attachment lugs were moved backward to the rear of the water jacket.

Batteries attached to the *Jäger-Bataillone* serving the cavalry units were increased from four to six guns in 1903. Commanded by an officer ranking as *Hauptmann* (Captain), each individual post-1903 machine-gun battery consisted of three sections of two guns, each directed by a subaltern. There were also thirteen NCOs, sixty-three men and no fewer than fifty-four horses – eighteen 'saddle horses' and thirty-six to pull the limbers and wagons. The officers and NCOs were armed with swords and handguns; the men carried Mauser carbines.

Experience with the Maxims showed that they were durable and efficient, and the value of the machine-gun in twentieth-century warfare was confirmed by German observers attached to the Japanese Army during the Russo-Japanese War of

An MG01 being fired in its lowest position. Note the quadrant-type elevator.

A typical horse-drawn limber and ammunition wagon.

1904–05. Infantry machine-guns and armoured machine-gun carts were tested in 1905–06, when sixteen six-gun batteries served the infantry participating in the annual *Kaisermanöver*.

The first *Maschinengewehr Abteilungen* (independent infantry machine-gun detachments) were formed in April 1906, each comprising six line and one reserve Maxims, four officers and eighty-seven men. These detachments were reorganized as machine-gun companies with effect from 23 May 1907. At this time, the monthly production capacity of the new Deutsche Waffen- und Munitionsfabriken factory in the Wittenau district of Berlin was rated at 700 Mauser rifles and 120 Maxim machine-guns.

A large appropriation in the 1908–09 defence budget enabled trials to find an ideal machine-gun

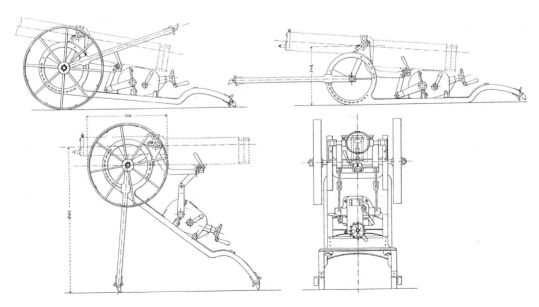

The MG03 sled, with a better elevating system and optional wheels.

49

to be completed. Many weapons were tried, but the reputation of the Maxim, solid and durable, prevented rival designs making much headway in Germany. Although the Schwarzlose became the regulation heavy machine-gun of the Austro-Hungarian, Dutch and Swedish armies, the Bergmann and the Dreyse designs were appreciably less successful.

A lightened Maxim was adopted as the *Maschinengewehr Modell* 1908 (MG08). It weighed 16.5kg, compared with the 26kg of the preceding patterns, and the weight of the sled mount was reduced from the 56kg of the *Schlitten* 03 to only 24kg for the new *Schlitten* 08. Experimental prismatic optical sights made by Zeiss were also tested, though the perfected *Zielfernrohr für Maschinengewehr* (ZFM), or 'Modell 1912', was apparently the work of Goerz.

The *Maschinengewehr* 08 (described in Part Two) was the principal small-calibre support weapon used by the German forces during World War One. However, it was not particularly portable and was accompanied by a Train of surprising proportions. The *Felddienst Ordnung* (Field Service Regulations) of 1908 noted that the transport section for each battery consisted of eleven horses for front-line duties, a four-horse stores wagon, a two-horse baggage wagon, a two-horse supply wagon, and a two-horse forage cart.

By 1912, machine-guns had been distributed to each infantry regiment in the form of six guns grouped in a special thirteenth company. The guns were issued in pairs to each of three sections, though – in peacetime, at least – one or two obsolete weapons were customarily available for training purposes.

AUTOMATIC RIFLES

Most of the semi-automatic rifles made prior to World War One were unnecessarily complicated, and the German designs were no exception. Even though they were backed by the resources of one of the world's largest and most successful gunmaking businesses, and were bolstered by the eagerness of the German Army to undertake field testing, the Mauser designers had little to show for nearly twenty years of work when World War One began.

The chronology of the auto-loading Mauser rifles is difficult to unravel, but the earliest seems to have been the C98. Patented in 1898–99, this was a long-recoil design locked by struts contained in a prominent housing at the front of the receiver.

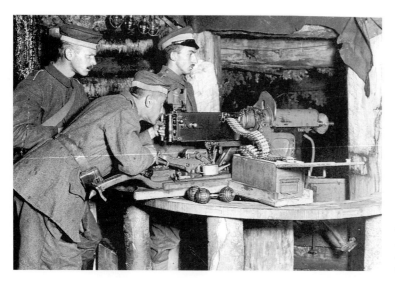

This photograph – obviously posed – shows an MG08 and Schlitten *08 in a sturdy defensive emplacement. Note the water-jacket shield and optical sight.*

When the gun was fired, the barrel ran back inside its casing until, after travelling about 8cm, the struts were cammed out of engagement. The barrel was stopped, and the bolt ran back alone to cock the internal hammer. A spring in the left side of the receiver returned the bolt to battery, stripping a new round into the chamber as it did so. The bolt rejoined the barrel, the barrel/receiver assembly moved forward, and the struts were cammed back into engagement to lock the mechanism ready for the next shot.

The patent shows a magazine case formed integrally with the elongated trigger-guard, a bolt-type charging slide on the right side of the receiver, and a prominent radial 'wing' safety-catch on the back of the breech cover. However, a rifle pictured in R.H. Korn's *Mauser-Gewehre und Mauser-Patente* has an internal magazine and a small oval trigger guard. A company of infantrymen tested guns of this type in 1901, and it is assumed that several hundred were made.

The clumsy C02, protected by a patent granted in November 1902, replaced the unsuccessful C98. Once the gun had fired, the barrel and bolt recoiled across the magazine well. A rapid-pitch thread in the multi-piece bolt rotated the locking lugs out of engagement near the end of the backward stroke,

and the bolt was held back as springs drove the barrel back to its original position. The final movement of the barrel activated the ejector, then tripped the retaining latch to allow the bolt to run back into battery, reloading the chamber and re-locking the lugs as it did so.

Although the C02 extracted more efficiently than its predecessor, it was needlessly complicated and the excessive mass of the moving parts made accurate shooting difficult. Revisions were being patented as late as 1905, but the project was abandoned shortly afterwards even though guns are occasionally reported with three-digit numbers.

The original patent shows a projecting magazine case and trigger-guard, but survivors customarily have internal magazines and small oval trigger-guards. A folding charging handle lay on the right side of the breech.

The C02 was replaced by the 1906-patent C06/08, a short-recoil pattern that was also offered as a handgun. The barrel and receiver slid back in the frame just far enough to release the bolt, which was locked either with lateral struts – cammed into the receiver wall behind the magazine well during recoil – or by a pivoting saddle.

The rifles were usually stocked similarly to the *Gewehr* 98, but had tangent-leaf backsights.

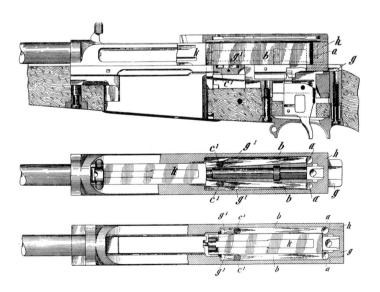

Typical of the auto-loading Mauser rifles developed prior to 1914, these patent drawings show one of the flap-locking patterns. Note the two pivoting bars locking in the sides of the receiver behind the magazine well.

Internal five-round magazines were standard, but twenty-round boxes were also made. Flap-lock guns have been reported with numbers as high as fifty, but the saddle types are much rarer.

The C10/13 rifle relied on a form of delayed blowback. When the gun was fired, recoil forced the entire gun backward against the firer's shoulder. However, an inertia block on top of the breech moved forward in relation to the remainder of the gun – a theory which had already been put to good use in the Sjögren shotgun – and cam slots cut in its top surface pivoted the locking flaps outward into the receiver walls. Deprived of support, the bolt then ran back to compress the telescoping mainspring unit.

When the mainspring returned the bolt, a new cartridge was stripped out of the magazine, a spring pushed the inertia block back to its rest position, and the cam slots moved the locking struts back into engagement.

The action was cumbersome and unreliable, as the slender locking bars were particularly prone to breakage, but Mauser designers were still doggedly persisting with it when World War One began.

THE FIRST SHORT RIFLES

The *Gewehr* 98 proved to be one of the world's most successful military rifles, but was by no means flawless. The receiver could crack below the thumb-clearance cutaway milled in the left receiver wall, in front of the bridge, but a more serious drawback was its great length and weight – as many German soldiers found to their cost in the trenches.

A few years after the *Gewehr* 98 had been adopted, the British began to issue the Rifle, Short, Magazine Lee-Enfield (SMLE). This was long enough to be an adequate infantry rifle, short enough to be an acceptable cavalry carbine, and destined to prove very handy in the trenches. The US Army had followed the British lead by accepting the 1903-pattern Springfield Rifle, which, ironically, was a slightly modified Mauser. The US government had even purchased rights to seven relevant patents.

The Germans toyed with *Einheitswaffen* (short rifles for universal issue), developing carbines of various lengths before concluding that the advantages they offered were largely illusory; recoil was too great, and the muzzle flash was often excessive. Large-scale distribution of *Gewehre* 98 was just beginning, and the Germans did not wish to shorten the rifle-bayonet combination in case of a war with France. The French had a long rifle with a long épée bayonet, and German troops would be at a severe disadvantage in a bayonet charge if a short rifle were substituted.

The full-length *Gewehr* 98 was retained for the infantry and riflemen, but the *Gewehr-Prüfungs-Kommission* had submitted a carbine to the *Kriegsministerium* by April 1900. A few pre-production examples were made in the Erfurt small arms factory and issued for trials with two cavalry squadrons and a foot artillery company.

The production pattern was not finalized until June 1900 and, although about 3,000 guns were made in Erfurt in 1900–01, the first *Kavallerie-und Artillerie-Karabiner* 98 was not successful. Replaced by the *Karabiner* 98 A in 1902, surviving carbines of the original type were withdrawn for conversion into *Zielkarabiner* (practice carbines) firing 5mm primer-propellant cartridges.

The *Karabiner* 98 was 945mm long, had a 435mm barrel, and weighed 3.325kg without its sling. The Lange-pattern backsight was graduated from 200m to 1,200m. The carbines worked similarly to the *Gewehr* 98 and shared the same butt profile, unlike the modified form of the later *Karabiner* 98 AZ.

The barrel and fore-end were much shorter than the rifle types; a special nose-cap protected the front sight; a stacking device was added beneath the muzzle; the turned-down bolt-handle took spatulate form; and a special small *Lange Visier* was fitted. The stock extended virtually to the muzzle, with a walnut hand guard running the entire length of the barrel from the nose-cap to the backsight. A sling bar was added on the left side of the barrel band and a sling aperture was cut through the butt, requiring the marking disc to be moved up and back towards the butt-plate.

The carbines were marked in much the same way as the *Gewehre* 98, except that the chamber mark displayed a crown over 'ERFURT/1900' (or '1901') and the designation stamp on the left side of the receiver may have read simply '98'.

Many complaints were voiced about the *Karabiner* 98, and the original pattern was abandoned in favour of the *Karabiner* 98 *mit Aufplanzvorrichtung für das Seitengewehr* 98 (1898-model carbine with attachment for the 1898-model bayonet), or Kar. 98 A, on 26 February 1902.

The new gun shared the dimensions and operating characteristics of its predecessor, but weighed about 3.425kg. The maximum setting of its original *Lange Visier* is believed to have remained at 1,200m. The carbines chambered the *Patrone* 88 when they were introduced in 1902, but experiments to adapt them to the more powerful S-Munition began in 1904.

The results were an altered chamber, deepened rifling grooves and a Lange *Visier* graduated from 300m to 1,800m. By the beginning of 1905, however, the *Gewehr-Prüfungs-Kommission* and the *Kriegsministerium* were expressing doubts as to the efficacy of converting the existing carbines for the S-Patrone, owing to fierce recoil, unpleasant muzzle blast, and a fearful muzzle flash. Production of the Kar. 98 A ceased, and the *Gewehr-Prüfungs-Kommission* developed a new weapon with a longer barrel: the *Karabiner* 98 AZ of 1908 (described in greater detail in Part Two).

Surviving *Karabiner* 98 A were withdrawn in 1908–09 for conversion to *Zielkarabiner*. Some had seen active service with the *Kaiserliche Schutztruppen* (German colonial forces in southern Africa), but those that remained in Germany were used for little more than extended troop trials.

About 6,000 *Karabiner* 98 A were made in the Erfurt small arms factory in 1902–05. They were identical mechanically with the *Gewehr* 98, though the spatulate bolt-handles were most distinctive. However, though the sling aperture through the butt and the sling ring on the left side of the barrel band were retained, a wooden hand guard stretched from the backsight base to a special barrel band/nose-cap unit. A bayonet bar and a half-length cleaning rod projected ahead of the nose-cap, below a narrow wooden fore-end stretching to the muzzle. Projecting ears on the muzzle block protected the front sight from the inside surface of the saddle boot.

Butt-discs were abandoned, unit marks being struck into the top of the butt-plate. The carbines inevitably bore their manufacturer's markings (for example, 'crown/ERFURT/1903') above the chamber, with the designation stamp on the left side of the receiver in front of the thumb-clearance cutaway. The designation originally read '98', but was subsequently changed to 'KAR.98'.

OTHER MACHINE-GUNS

Although the Deutsche Waffen- und Munitionsfabriken-made Maxim was by far the most successful German machine-gun made prior to 1914, rival designs were also promoted from time to time. Among them was the Bergmann, patented in the name of Theodor Bergmann in 1901 even though the design was actually the work of Louis Schmeisser.

The earliest prototype was a crude device mounted on a tripod made largely of wood, but it worked well enough to allow development of a better gun to continue.

The recoil-operated Bergmann-MG02 was locked by a rising block, housed in the barrel extension, which engaged in the recess in the top surface of the bolt. The drive for the articulated metallic 'push-through' belt was taken from the movement of the breech-bolt, and a hinged top-cover gave ready access to the feed way. The conventional tube-leg tripod mount had provision for elevation and traverse, and a sprung bicycle-type seat was added for the gunner.

The Bergmann-MG02 was easily recognizable by virtue of its unusually slender receiver and the large folding-bar sight on the receiver top. Water-cooled, it fed from right to left – often with the aid of a large belt-box attached to a bracket on the receiver-side.

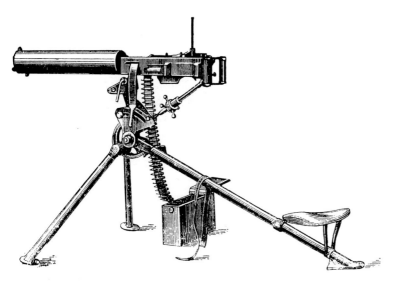

By the standards of the day, the Bergmann-MG02 was an advanced design that deserved a better fate. Part of its under-achievement was, however, due entirely to a loss of production facilities. Bergmann made a wide variety of goods, ranging from railway signals and ornamental metalwork to bicycles and automobiles. However, the auto-loading pistols designed by Schmeisser and marketed under the Bergmann name were made by V.C. Schilling of Suhl under a sub-contract arrangement. Schilling was purchased by Sempert & Krieghoff in 1904 and the agreement with Bergmanns Industriewerke was subsequently terminated.

Bergmann sold the rights to the 'Mars' pistol to Anciens Établissements Pieper of Herstal, but work on machine-guns seems to have been terminated. Interest in the project was renewed in 1908 in the Bergmann 'Abteilung Waffenbau' in Suhl, perhaps coinciding with the competitive tests that eventually led to the adoption of the MG08.

It is assumed that remedial work on the Bergmann-MG02 – though clearly based on the original patent – was entrusted to Hugo Schmeisser, son of Louis, who had remained with Bergmann after his father's departure to work for Rheinische Metallwaaren- und Maschinenfabrik of Sömmerda (better known by the wartime acronym 'Rheinmetall'). The elder Schmeisser subsequently designed the Dreyse machine-gun (*see* Part Three, '1915: Germany').

THE RISE OF THE LUGER

The Borchardt-Luger, later known as the 'Parabellum' (or 'Luger' in English-speaking countries), was lighter, handier and more efficient than the Mauser C96. However, the *Gewehr-Prüfungs-Kommission* was worried by its complexity, lack of power in the original 7.65mm cartridge, and complaints that it was difficult to tell whether the gun was loaded or cocked. Intriguingly, the *Gewehr-Prüfungs-Kommission* claimed to have made 'several improvements in order to prevent vital parts breaking and to increase the reliability of the pistol'.

Another Mannlicher was submitted in December 1901 and, finally, a third series of field trials began in April 1902 with fifty-five improved six- and ten-shot Mauser-Pistolen C96; a similar quantity of Borchardt-Lugers, forty of which lacked grip safeties (*ohne automatische Sicherung*); and fifteen Mannlichers to be tested as an officer's weapon.

Criticism of the stopping power of the standard 7.65mm bullet inspired development of the first

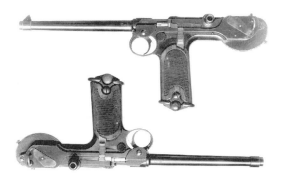

The Borchardt pistol, patented in 1893, was the prototype of Borchardt-Luger or Parabellum.

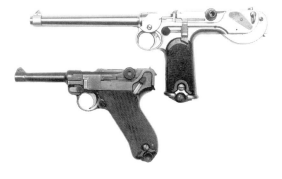

A comparative view of the Borchardt (top) and its successor, the Parabellum. Colonel W. Reid Betz

9mm Parabellum cartridge, which was made simply by enlarging the neck of the 7.65mm type. The result was DWM case 480, which had a slight bottleneck. By the summer of 1902, however, Georg Luger had produced a straight-wall case – DWM 480C – which shared the head dimensions of the original 7.65mm Parabellum and allowed pistols to be chambered for either round simply by changing barrels.

On 18 December 1902, Vickers, Sons & Maxim informed the British Director-General of Ordnance that a 9mm-calibre pistol 'could be submitted for trials in the third week in January next [1903] …' This is the earliest authenticated mention of 9mm pistols. A Bavarian observer reported that experiments were underway with bullets 'of larger calibre' and 'different types of … material' in March 1903; in December, the *Gewehr-Prüfungs-Kommission* told the *Reichsministerium* that the Borchardt-Luger was preferred even though it lacked a loaded-chamber indicator and the toggles sometimes failed to close.

Tests were still underway in March 1904 to improve 'wounding efficiency' by increasing the calibre or by using bullets of special material. In May, however, the *Gewehr-Prüfungs-Kommission* acknowledged delivery of an 'improved Luger-pistol (Parabellum), calibre 9mm … with a flat-nose bullet'. Although the reports provide few clues to these changes, they are believed to refer to

the addition of a combination extractor/loaded-chamber indicator. A patent for this device had already been sought.

In June 1904, the *Gewehr-Prüfungs-Kommission* reported that the Luger had been improved sufficiently to justify adoption. However, Mauser was well aware of the preference for a rival design and intended to 'submit his improved self-loading pistol in a larger calibre in August'.

Adoption of the Luger was temporarily deferred while improved Mauser, Frommer and Vitali pistols were tried. While the German Army procrastinated, the *Kaiserliche Marine* began handgun trials of its own when, on 1 August 1904, the *Reichs-Marine-Amt* (Imperial Navy Office)

The original 1900-pattern Borchardt-Luger or Parabellum with a 120mm barrel, a grip safety mechanism, and distinctively cutaway toggle grips.

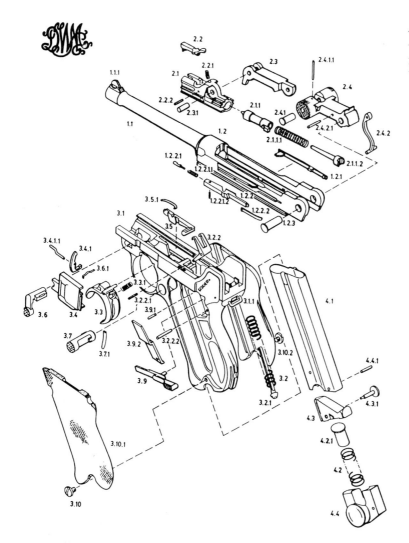

An exploded view of the 'New Model' or coil-spring pattern of the Parabellum or Luger pistol.
Reinhard Kornmayer

ordered the issue of five '9mm *Selbstlade-Pistolen Modell* 1904' to the Baltic navy station. Each gun was to be accompanied by a holster-stock, three magazines, a cleaning rod, a screwdriver and three dummy cartridges.

Trials were to be undertaken by land-based naval units, by ships of the active battle fleet, and aboard the gunnery test ship. The goal was a replacement for the antiquated 1883-type revolvers in the hands of landing-party personnel who did not carry rifles – officers, petty officers, signalmen, engineers and stretcher-bearers.

Reports submitted by September 1904 agreed that the pistol was not only far superior to the revolvers, but also much handier than the rifles. It seems as though the *Reichs-Marine-Amt* had already asked the *Reichsministerium* to order 2,000 pistols from Deutsche Waffen- und Munitionsfabriken of Berlin.

Eighteen semi-experimental guns were issued in August 1905 to personnel of the East African Expeditionary Force, despatched to crush the Maji-Maji rebellion in German East Africa, and a similar quantity went to South West Africa, where the Herero-Nama revolt was being quelled.

Unfortunately, the sailors had expressed a preference for rifles carried by native bearers. The officers purchased smaller pistols for self-defence, as the excessive weight of the Parabellum and the awkwardness of the grip safety were heartily disliked.

Although many millions of words have been written about the Luger pistol, doubt still clouds the key elements of its development history. The Deutsche Waffen- und Munitionsfabriken factory designation 'Modell 1904' referred not to the earliest guns issued to the *Kaiserliche Marine* – which had a toggle-lock and a riband mainspring – but instead to the developmental phase of what is now usually known as the 'New' or '1906 Model'.

The replacement of the riband mainspring with a coil pattern still excites much controversy. European enthusiasts, in particular, have been fighting a verbal battle since claims were made by the late Dick Deibel that the coil-spring was the work of Dutchmen.

The Dutch ordered (but did not receive) coil-spring guns in November 1904, and the Swiss authorities learned of the existence of 'New Model' coil-spring guns late in 1905. Thus the design of the 'New Model' pistol had clearly been perfected by the end of 1905, even though series production had yet to begin, and features such as the coil-spring and the combination extractor/ loaded-chamber indicator were undoubtedly essential to its design.

No patent was sought for the coil-spring, however, as the toggle action was unchanged and the Parabellum was scarcely the first pistol to incorporate coil-springs. The *Gewehr-Prüfungs-Kommission* and the Dutch *Artillerie-Inrichtingen* may both have suggested simply that a 'better' spring was required (allowing both to claim credit), though detailed recommendations need not have been made. Virtually every military trial board drew attention to the 'weak closure' of the toggle system, but none offered ways of improving it.

A claim has also been made on behalf of Adolf Fischer (1868–1938), a junior officer in a Bavarian infantry regiment when he developed two straight-pull rifle actions in the 1890s. This work brought Fischer to the attention of the Bavarian *Kriegsministerium*. He offered a semi-automatic pistol in 1899 and was seconded to serve with the *Gewehr-Prüfungs-Kommission*, where he stayed until 1908.

In a letter to his departmental head, written on 24 October 1904, Fischer claimed that 'the suggestions made by me to improve on the original design [of the Luger] have been forwarded by the Commission to the Ministry with a request to be approved'. Although this has been used to 'prove'

One of the first 9mm-calibre Parabellums to be tested by the German Army. Dr Rolf Gminder

that Fischer was responsible for the coil-spring, there is nothing to substantiate claims which were repeated in recollections published in 1929–30 in *Artilleristische Rundschau* and *Zeitschrift für das gesamte Schiess- und Sprengstoffwesen.*

POSTSCRIPT

When World War One began, the German Army was among the best-equipped in Europe. Virtually every front-line unit was armed with the Maxim machine-gun, the Mauser rifle or the Parabellum pistol, and many of the guns held in reserve would have been the envy of front-line regiments elsewhere.

The issue of these weapons had been made possible only by the zeal with which successive governments had linked exports and the arms trade, ensuring that practically every country that accepted German engineering assistance – to build railways, for example – was also pressured into taking state-subsidised German weapons. Companies such as Krupp, Mauser and Deutsche Waffen- und Munitionsfabriken, therefore, numbered among the largest and most successful in the world.

Though the Mannlicher had enjoyed considerable popularity prior to 1895, the introspective Austro-Hungarians gave Österreichische Waffenfabriks-Gesellschaft so little support that the company was forced to enter a Mauser rifle-making cartel with Waffenfabrik Mauser, Deutsche Waffen- und Munitionsfabriken and Fabrique Nationale d'Armes de Guerre at the beginning 1897.

Such was the robustness of the German export drive, helped by the emergence of the 1893-pattern rifle, that virtually every army in Central and South America was armed with Mausers by 1914.

Sales of these rifles also encouraged purchases of Deutsche Waffen- und Munitionsfabriken-made Maxim machine-guns, though the trade in handguns was comparatively small as enthusiasm was muted by complexity and high price. The Parabellum often found less favour than the Mauser C96, but neither sold in vast numbers prior to World War One.

Sailors of II. Torpedo-Division and their instructors pose in Wilhelmshaven on completion of basic training, 1911. They are armed with 1904-type navy Parabellums and bayonets. Hartmut Kordeck

4   Turkey

By the last quarter of the nineteenth century, the heyday of the Ottoman Empire was long past. A lengthy period of decline had been accompanied by substantial losses of territory, interrupted only by a few paltry gains from Russia after the Crimean War. Large tracts of the Balkans had gone immediately after the Russo-Turkish War of 1877–78, and Egypt had become a British protectorate in 1882. By 1914, Turkey was regarded with justification as the 'Sick Man of Europe' and was destined to lose even more land at the close of hostilities in 1918.

Comparatively small in peacetime, the Turkish armed forces were greatly enlarged during World War One. Despite the efforts of German advisers, however, ineffectual leadership all too often masked the courage and undeniable fortitude of the Turkish soldier.

The history of Turkish military firearms from the close of the Crimean War to the final capitulation on 31 October 1918 can be divided into three periods: a reliance on obsolete rifle-muskets, a breech-loading era satisfied largely with American weapons, and finally the German domination which lasted until the end of World War One.

Although Turkey had ordered 60,000 Remingtons for service in Egypt in the late 1860s and 50,000 1866-type Winchesters in 1870–71, the first large-scale orders – for an amazing 600,000 Peabody-Martini rifles – were placed with the Providence Tool Company in the summer of 1872. These guns were supplied in batches until 1881, when a supplementary order for 100,000 was placed, and deliveries continued until the Providence Tool Company failed in 1885.

An inventory taken in Turkey at the beginning of the Russo-Turkish War in 1877 revealed about 310,000 Peabody-Martini rifles and carbines; 323,000 Sniders, converted from Enfield and Springfield rifle-muskets; 39,000 Winchesters; and about 20,000 revolvers, mostly 'Russian Model' Smith & Wessons and similar copies supplied by Ludwig Loewe & Cie.

An assessment made some twenty years later, in the autumn of 1896, showed that there were 220,000 1887-pattern 9.5mm Mauser rifles and 4,000 carbines; 481,900 M1890 and M1893 7.65mm Mauser rifles; about 500,000 .45-calibre Peabody-Martini and Martini-Henry rifles; 150,000 Sniders with a nominal calibre of 12mm; 50,000 .44-calibre Winchesters; and a similar number of 11mm Remingtons.

Mauser rifles continued to be purchased in large numbers, but the procurement of weapons could depend on the whim of the Sultan. In the mid-1890s, after nearly a decade of reliance on Mausers, the Sultan suddenly decided to arm his cavalrymen with the Krag-Jørgensen! Only a handful of these Steyr-made guns ever appeared for trials, however, and it was some years before the cavalry received new Mausers.

Constant bickering and skirmishes with neighbouring states, many of which eyed Turkish territory enviously, fuelled a constant need for new weapons. A short, but full-scale war was fought with Greece in 1897 and friction continued into the twentieth century. The Balkan Wars of 1912–13 also deprived Turkey of land.

When hostilities began in earnest in 1915, Turkey still relied on Germany for Mauser rifles,

A view of Istanbul, taken early in the twentieth century. The mixture of sail and steam shipping in the harbour typifies the stage that Turkish technology had reached. This was reflected in the small arms that they were using at the time.

Deutsche Waffen- und Munitionsfabriken-made Maxim machine-guns, and Krupp field guns. Ammunition had been supplied by Deutsche Waffen- und Munitionsfabriken, and bulk deliveries of propellant came from Vereinigte Köln-Rottweiler Pulverfabriken early in the twentieth century. Cartridge-making facilities were built with German help, the Turkish Army depended greatly on its German advisers, and even the Berlin-Baghdad railway – the main communications artery – had been built by German engineers with German money.

Much war *matériel* had been lost during the Balkan Wars, but sufficient modern rifles were available to equip the front-line units in 1914. This was partly due to a particularly astute move by the Turkish authorities, among the first to order Mauser-type magazine rifles in quantity, who insisted on a clause in the 1887 contract to force the substitution of a better Mauser rifle should one be adopted elsewhere.

The enormous contracts given to Waffenfabrik Mauser did much to ensure the company's long-term prosperity, and rapid strides in the design of small-bore magazine rifles in the 1890s ensured that the Turks ultimately received several different patterns.

Turkish rifles made prior to about 1926 can usually be identified by marks in Arabic; in addition, many of the more modern Mausers have a *Toughra* (a calligraphic symbol unique to each ruler) on the chamber-top.

PART TWO:
THE GUNS

5 Austria–Hungary

HANDGUNS

In addition to the regulation revolvers and pistols listed below, individual officers purchased pistols of their own. A few acquired Mannlichers, which are described briefly in the introduction, but the most popular purchase was the 1909-pattern Steyr – a 7.65mm-calibre blowback licensed from Pieper in Belgium. This is described in Part Three (*see* '1914') as a wartime impressment.

Austria–Hungary: Weights and Measures

Austria-Hungary used an indigenous system of weights and measures prior to the adoption of the metric system in 1873 (though metrication was not implemented until 1 January 1876).

Dimensions were usually expressed on the basis of the *Kaiserfuss* ('Imperial Foot'), also known as the *Wiener-fuss* ('Vienna foot'), which was divided into *zoll, linie* and *punkt*, each being a twelfth of its predecessor. The *Kaiserfuss* was the equivalent of 1.0373ft (12.45in or 316.2mm); a *zoll* was 1.037in, a *linie* was 0.0864in, and a *punkt* was 0.0072in.

Rifle sights were graduated in paces or *schritt*, each originally being equivalent to 28.8 'Wiener zoll'; a pace, therefore, was the equivalent of 30.32in or 77cm. Oddly, *schritt* graduations continued in use until 1918 even though the metric system was by then being used to control manufacturing tolerances.

Differing Imperial- or metric-measure equivalents are often given for the Austrian *schritt*, usually without realizing that there are two different dimensions: the original 77cm pattern and the 75cm (29.53in) *dezimal schritt* ('decimal pace') used from the mid-1880s onward.

Weights were customarily expressed in *pfünde*, (pounds), which were divided into four *vierling*, 16 *unze*, 32 *loth*, 128 *quentchen* or 512 *pfennige*. Only the *pfund* and the *loth* were customarily used for gun weights, giving figures such as '8.21/32'; these have been mistakenly transcribed directly into pounds avoirdupois, but the true equivalent of the Austrian *pfund* was 1.235lb or 560g.

The 7.65mm Steyr, a Belgian Pieper design, was a popular self-defence weapon amongst the officer class during World War One.

Front-Line Patterns

8mm Repetierpistole Modell *1907 (M7)*

The origins of this pistol remain something of a mystery. Some writers have identified the designer as an engineer named Wasa Theodorovic, but others have suggested that Theodorovic was simply a patent agent.

The original gun had a barrel which slid backward under recoil until it was stopped by a shoulder in the frame; a cam slot in the operating-slide head then rotated the bolt, unlocking it from the barrel, and the bolt ran back to the limit of the opening stroke. The return spring then pushed the parts back, stripping a new round into the chamber and rotating the locking lug back into engagement.

The Theodorovic pistol had an internal magazine, loaded from a six-round charger with a sliding thumb-piece; a trigger which was isolated from the operating mechanism; and a mechanical hold-open released by a sliding catch set into the left side of the frame above the grip. However, the gun was primitive and clumsy, and did not impress the *Militär-Technische-Komitee*.

Success awaited the substitution of a barrel-locking system developed by Karel Krnka. The subject of patents granted in 1894–1908, often in the name of Georg Roth, this relied on a barrel which was locked in a slide or 'sleeve' reciprocating within a

8mm *Repetierpistole Modell* 1907 (M7)	
Synonym:	Roth-Steyr M1907
Adoption date:	5 December 1907

Data taken from *Einrichtung und Verwendung der Repetierpistole M7* (Vienna, 1908)

Length:	243mm (9.57in)
Weight:	1,000g (35.3oz) empty
Barrel length:	128.3mm (5.05in)
Chambering:	8 × 18.7mm, rimless
Rifling type:	four-groove, concentric
Depth of grooves:	0.2mm (0.008in)
Width of grooves:	3.5mm (0.138in)
Pitch of rifling:	one turn in 250mm (9.84in), RH
Magazine type:	single-row internal box
Magazine capacity:	10 rounds
Loading system:	charger
Front sight:	open barleycorn
Backsight:	a fixed notch in the charger-guide housing
Velocity:	320m/sec (1,050ft/sec) at 12.5m (41ft)
Bullet weight:	7.5g (116gn)

The Repetierpistole *M7 or Roth Steyr.*

closed tubular frame. Krnka-Roth pistols were hawked around Europe at the beginning of the twentieth century – some were tested in Britain in 1902 – and soon attracted the attention of the Austro-Hungarian military authorities.

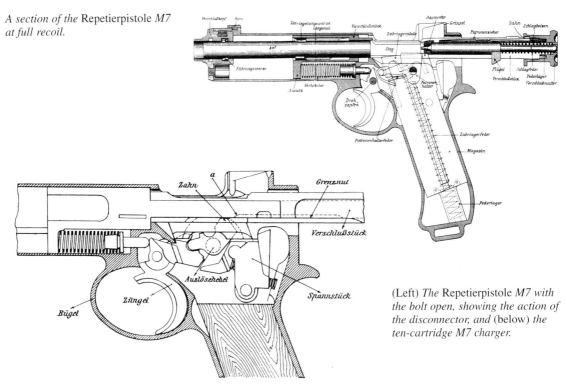

A section of the Repetierpistole *M7 at full recoil.*

(Left) *The* Repetierpistole *M7 with the bolt open, showing the action of the disconnector, and* (below) *the ten-cartridge M7 charger.*

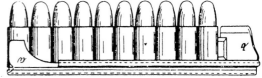

Tests with the 'Modell I' showed that improvements could be made, and three examples of the 'Modell II' or 1904-type were tested extensively by the *Militär-Technische-Komitee* throughout the summer and autumn of 1905. The Krnka-Roth pistols fired nearly 6,000 rounds in competition with 7.65mm Mannlichers. The weapons were re-examined in 1906, when the Krnka-Roth, though much less accurate, proved to be more durable than its rival. Field trials were undertaken in the spring of 1907, a few last-minute adjustments were made, and the *Repetierpistole* M7 was adopted to replace the revolvers then in the hands of the cavalry and mounted artillery.

The M7 fired an 8mm centre-fire cartridge loaded with a steel-jacketed lead core bullet weighing about 7.5g. The rimless straight-wall case, made of brass, was 18.75mm long and contained 0.28g of M97/C flake nitro-cellulose propellant. The loaded rounds were 29mm long and weighed 10.6g Ten

were loaded in sheet-steel chargers and issued in cardboard packets, total weight being about 120g. Markings on the cartridge-packet labels, which were printed in black, read '10 St. *8mm* M. 7', 'für Repetierpistolen' and 'schf. Patronen' ('ball cartridges') together with the name of the assembler and the month and year of assembly – for example, 'Lichtenwörth III 1909'. Five thousand cartridges were packed in a cartridge box, identified by a yellow label printed in black with additional details of the suppliers of the cases, bullets and propellant.

The pistols were made by Österreichische Waffenfabriks-Gesellschaft in Steyr and by Fegyver és Gepgyár in Budapest, 'WAFFENFABRIK STEYR' or 'FEGYVERGYÁR BUDAPEST' appearing on the top surface of the barrel casing. The guns also

---

8mm *Repetierpistole Modell* **1907 (M7)**

Internal Arrangements

When the gun is fired, the barrel and slide run back about 8mm until two small trapezoidal lugs on the barrel – immediately behind the muzzle – are rotated by cam tracks in the muzzle bush. This releases the locking lugs on the barrel from their recesses in the slide. The barrel is stopped by a lug on the frame, allowing the slide to run back out of the frame alone. The mainspring then returns the components to their rest position, reloading the chamber from the internal magazine as it does so.

The special double-action trigger mechanism completes the retraction of the striker immediately before firing, which provided cavalrymen with a safety feature by preventing a single-action hammer being jolted off the sear accidentally.

External Appearance

The Roth-Steyr has a distinctive full-length tubular barrel shroud and prominent charger guides milled in a block on top of the receiver. The charger guides double as the backsight, whereas the front sight is mounted on a complicated machining containing the muzzle bush.

The walnut grips have diagonal grooves, a fixed lanyard loop lies on the base of the butt, and unit markings may be struck into the special serrated-edge washer that anchors the grip bolts. A hold-open catch and a cartridge-retaining latch, set into the left side of the frame, have chequered heads.

Operation

When the gun fires, the action automatically extracts and ejects the spent case, then reloads the chamber. If the magazine is empty, however, the magazine follower holds the bolt to the rear. A new charger is placed in the guides on top of the breech and the ten cartridges are forced down with the thumb as far as they will go. As the charger is tugged free of the guides, the bolt runs forward to chamber the uppermost round. The tip of the firing pin protrudes from the back of the cocking-piece to show not only that the mechanism is loaded, but also that the trigger is at half cock. A pull on the double-action trigger is sufficient to fire the gun.

The magazine can be unloaded by retracting the cocking-piece as far as it will go and pressing the hold-open catch to lock it open. Pushing down on the cartridge-retainer latch directly above the left grip expels any cartridges remaining in the magazine. A slight rearward pull on the cocking-piece disengages the hold-open catch to leave the bolt held back by the magazine follower, and another press of the cartridge-retainer closes the action on an empty chamber.

---

customarily bore acceptance marks such as 'W-n [eagle] 11' on the back of the frames.

The first deliveries – from Steyr – seem to have been made in 1909. Field-trials guns apparently had chequered grips and lacked marking discs. A few experimental guns were made with safety-catches, but the only major change made during the production life of the Roth-Steyr was the addition of a disconnector in the sear train to prevent 'doubling' (firing more than one shot for each press of the trigger). This occurred in 1909, after about 7,000 guns had been made, and can be identified by the addition of an axis pin visible on the right side of the frame above the grip.

Production is believed to have amounted to about 55,000 guns in Steyr (1909–13) and 35,000 in Budapest (1911–14) before work stopped in favour of the Steyr M12 and the Frommer 12M pistols respectively.

9mm Repetierpistole Modell *1912 (M12)*

This pistol was developed as a speculative venture in 1909–10 on the basis of the Roth-Steyr, though Karel Krnka is said to have been disgusted with the changes. The basic work is usually attributed to an Österreichische Waffenfabriks-Gesellschaft engineer named Karl Murgthaler, but the detailed development was the work of Helmut Bachner and

The Repetierpistole *M12 or Steyr-Hahn. This is a Chilean contract gun: note the marks on the slide.*

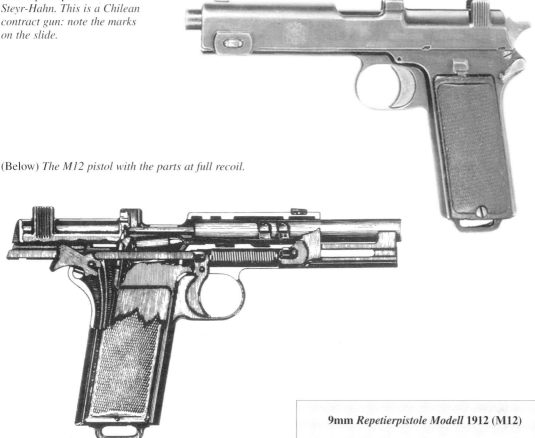

(Below) *The M12 pistol with the parts at full recoil.*

9mm *Repetierpistole Modell* 1912 (M12)	
Synonym:	Steyr-Hahn
Adoption date:	October 1914(?)
Length:	217mm (8.54in)
Weight:	1,080g (38.1oz) loaded
Barrel length:	129.5mm (5.1in)
Chambering:	9 × 22.7mm, rimless
Rifling type:	four-groove, concentric
Depth of grooves:	0.2mm (0.008in)
Width of grooves:	4.4mm (0.175in)
Pitch of rifling:	one turn in 200mm (7.87in), RH
Magazine type:	single-row internal box
Magazine capacity:	8 rounds
Loading system:	charger or single rounds
Front sight:	open barleycorn
Backsight:	fixed 'V'-notch
Muzzle velocity:	335m/sec (1,100ft/sec)
Bullet weight:	7.5g (116gn)

Adolf Jungmayr. Patented in Britain in 1911–12, the Steyr-Hahn was offered commercially as the Modell 1911.

It was purchased in small quantities by Chile (as the Mo. 1911), and then by Romania (Md 1912) before being adopted by Austria–Hungary shortly after World War One began. However, tests had obviously been underway for some time, as gun No. 01, cased with Steyr pocket and personal-defence pistols, was presented to Emperor Franz Josef on 18 August 1913.

Chile's guns can be identified by 'EJÉRCITO DE CHILE' on the right side of the slide and the national Arms (a star on a halved shield supported

---

9mm *Repetierpistole Modell* 1912 (M12)

Internal Arrangements

Though retaining recoil operation and a rotating-barrel lock, the Steyr-Hahn is much less complicated than the preceding Roth-Steyr. When the gun fires, the slide and the barrel run back, locked together, until a lug on the underside of the barrel – working in a cam track in the frame – rotates the locking lugs through 60 degrees to disengage the slide. The barrel halts when a fourth lug strikes an abutment on the frame, leaving the slide to run back alone. The mainspring then returns the slide from the limit of its backward travel, stripping a new round into the chamber from the magazine. The last stages of the movement back into battery turn the locking lugs back into engagement with the slide, and the gun can be fired again. The disconnector is combined with the ejector, and a conventional single-action trigger system is fitted. An external hammer is cocked each time the slide runs back.

External Appearance

The M12 is easily distinguished by the retraction grips on the rear of the slide, by prominent charger guides on top of the breech behind the ejection port, and by the unusually square angle between the grip and the bore-axis. The magazine is internal, but a cartridge-release button protrudes from the top of the left grip; the grips are diced walnut. A radial safety lever on the rear left of the frame can also be used to lock the slide back, and a lanyard loop is fixed to the base of the butt.

Operation

The M12 works semi-automatically until the magazine follower holds the slide open after the last round has been ejected, showing the firer that he has an empty gun. The slide is then retracted far enough to be held by turning the tip of the safety lever up into the appropriate notch. Cartridges are inserted singly into the magazine or, alternatively, a charger is placed in the guides and eight rounds are stripped down into the magazine in a single movement. When the charger has been removed, pressing the safety lever head releases the slide to run forward and chamber a new round.

Emptying the magazine is simply a matter of holding back the slide with the safety lever, then pressing the cartridge-retainer latch. This expels the cartridges – throwing them several feet – and the slide can be returned to battery.

---

by an Andean deer and a condor) on the left, whereas Romania's guns have a large crown above 'Md 1912' on the slide.

A pre-production series of at least 100 guns was made in 1911. The slide-retaining block housing at the front of the side, which was much larger than the perfected pattern, distinguished these. This change had already been made by the time of the Chilean contract. The head of the safety-catch was altered at much the same time, but the only other change to be made prior to 1918 concerned the follower on the sheet-steel charger, which was enlarged from the end of 1915.

The powerful 9mm M12 cartridge consisted of a steel-jacketed lead bullet loaded into a straight rimless brass case containing nitro-cellulose flake propellant. Two loaded eight-round chargers were packed in a small cardboard box, labelled '16 St. 9*mm* M. 12', 'für Repetierpistolen' and 'schf. Patronen', together with the place and date of assembly.

Pistols customarily bore nothing other than 'STEYR' and the date on the rear left side of the slide, together with a large 'S'. Acceptance marks – for example, 'W-n [eagle] 15' – lay on the front left side of the trigger-guard bow. The serial numbers took a cyclical form of 10,000-gun blocks distinguished by suffix letters. It is believed that about 262,000 were made prior to the Armistice. However, the Austro-Hungarians also recaptured some of the contract pistols supplied in 1913–14 to Romania; dimensionally identical with the standard M12, they bore a large crown on the front left side of the slide above 'Md 1912'.

7.65mm Pisztoly *12M*

This gun was the culmination of more than a decade of work by Rudolf Frommer (1867–1936). Often promoted by Georg Roth, the earliest Frommer pistols may be confused with similar-looking Krnka designs, though most of them utilize long recoil to operate the breech-lock; Krnka patterns generally rely on short recoil.

7.65mm *Pisztoly* 12M

Synonym:	Frommer-'Stop'
Adoption date:	1912
Length:	165mm (6.5in)
Weight:	610g (21.5oz) empty
Barrel length:	96.5mm (3.8in)
Chambering:	7.65 × 17mm, semi-rim
Rifling type:	four-groove, concentric
Depth of grooves:	0.17mm (0.007in)
Width of grooves:	2.9mm (0.114in)
Pitch of rifling:	one turn in 240mm (9.45in), RH
Magazine type:	single-row detachable box
Magazine capacity:	7 rounds
Loading system:	replacement magazine
Front sight:	open barleycorn
Backsight:	fixed 'V'-notch
Muzzle velocity:	305m/sec (1,000ft/sec)
Bullet weight:	5g (77gn)

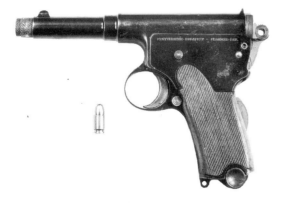

The 1910-type Frommer pistol was made in small quantities.

7.65mm *Pisztoly* 12M

Internal Arrangements

The gun is a refinement of the 1910 Frommer, differing principally in the position of the springs. When the 12M fires, the bolt and the barrel recoil inside the frame, locked together, for considerably greater than the length of the cartridge case. The initial movement compresses the barrel-return spring in its chamber in the top of the frame. By the time the parts reach the limit of their backward travel, the bolt has been rotated to disengage lugs on the bolt head from the barrel extension; spring pressure then returns the barrel while the bolt is held to the rear. The spent case is extracted and ejected through a port on the right side of the breech. The bolt is tripped as the barrel comes to a halt, returning under spring pressure to reload the chamber and rotate the locking lugs into position. Bolt and barrel can then run forward into battery at the end of the closing stroke. As the breech opens, the tail of the bolt cocks an external hammer.

External Appearance

The Frommer-'Stop' pistol would have an external affinity with the FN-Brownings had it not been for the chamber in the frame-top containing the barrel and bolt-return springs and their associated guide rod. Closer inspection reveals that the barrel casing is fixed, and that the moving parts recoil within it. A grip safety is set into the rear strap of the butt, the grips are chequered rubber with an 'FS' trademark, and a fixed lanyard loop lies on the butt-heel alongside the serrated magazine-release catch. Sights consist of a blade and a fixed 'V'-notch on top of the frame.

Patented in 1912, the Frommer-'Stop' was adopted as the standard pistol of the *Honvéd*, the Hungarian home-defence force, when production of the M7 Roth-Steyr (qv) stopped in the Fegyvergyár factory in Budapest. When the *k.u.k Armée* decided to replace the M7 Roth-Steyr with the M12 Steyr-Hahn, the *Honvéd* authorities decided that the Steyr-made gun was too large, too powerful and probably too expensive. Thus the Frommer-'Stop' was selected instead.

Part Two: The Guns

Chambered for the standard 7.65mm Automatic Pistol cartridge (.32 ACP), the guns were marked 'FEGYVERGYÁR–BUDAPEST • FROMMER–PAT. STOP Cal. 7,65m/m (.32)' on the left side of the frame alongside the spring chamber. At least 100,000 served during World War One, but post-1918 production and serial numbers that can exceed 330,000 complicate accurate assessments.

8mm *Armeerevolver Modell* **1898 (M98)**	
Synonym:	Rast & Gasser army revolver
Adoption date:	1898
Length:	225mm (8.86in)
Weight:	980g (34.6oz) empty
Barrel length:	116mm (4.57in)
Chambering:	8 × 27mm, rimmed
Rifling type:	four-groove, concentric
Depth of grooves:	0.2mm (0.008in)
Width of grooves:	2.8mm (0.11in)
Pitch of rifling:	one turn in 150mm (5.9in), RH
Magazine type:	rotating cylinder
Magazine capacity:	8 rounds
Loading system:	single rounds
Front sight:	open barleycorn
Backsight:	fixed notch
Muzzle velocity:	229m/sec (750ft/sec)
Bullet weight:	7.8g (120gn)

The M1898 Rast & Gasser revolver. Ian Hogg

Second-Rank Patterns

8mm Armeerevolver Modell *1898 (M98)*

Adopted to replace the earlier Gasser and Gasser-Kropatschek designs (qv), the M98 was the work of August Rast – later a partner in Rast & Gasser, successor to Leopold Gasser Waffenfabrik, the principal manufacturer.

The M98 was unremarkable mechanically, though it had Abadie-type lockwork, a rebounding hammer, and a firing pin mounted in the frame instead of forged as part of the hammer nose. The solid-top frame was much sturdier than the M1870, M1870/74 or M1878 patterns. Adopting a smaller cartridge allowed two chambers to be added in the cylinder, loading being undertaken by inserting cartridges singly through a swinging gate on the right side of the breech. Spent cases could be punched out with a sliding ejector rod, which projected from the right side of the barrel.

The trigger mechanism was a double-action design and the square-base grip projected almost at right angles to the axis of the bore. One of the most interesting features of the M98, however, was the ease with which it could be stripped for cleaning. Pulling down on the front of the trigger-guard released a swinging plate on the left side of the frame, pivoted behind the hammer, to give access to the firing mechanism.

The M98 was made in large numbers, as most of the 180,000 revolvers given to Italy as reparations after World War One were apparently of this type. The earliest were marked as the product of Gasser – apparently until 1903 – but later ones bore the 'Rast & Gasser' name.

Obsolete Patterns

9mm Offiziersrevolver Modell *1878 (M78)*

This improved version of the M1870/74 Gasser design, credited to Alfred von Kropatschek, does not seem to have been made in large numbers. Confined largely to infantry officers and gendarmerie units, the M1878 was distinguished by its small size and plain-surfaced five-chamber cylinder. The barrel

68

9mm *Offiziersrevolver Modell* 1878 (M78)

Synonym:	Gasser-Kropatschek revolver M1878
Adoption date:	September 1878(?)
Length:	229mm (9.02in)
Weight:	770g (27.2oz) empty
Barrel length:	122mm (4.8in)
Chambering:	9 × 26mm, rimmed
Rifling type:	four-groove, concentric
Depth of grooves:	0.2mm (0.008in)
Width of grooves:	3.25mm (0.128in)
Pitch of rifling:	one turn in 310mm (12.2in), RH
Magazine type:	rotating cylinder
Magazine capacity:	5 rounds
Loading system:	single rounds
Front sight:	open barleycorn
Backsight:	fixed notch
Muzzle velocity:	219m/sec (720ft/sec)
Bullet weight:	10.2g (157gn)

11mm *Armeerevolver Modell* 1870/74 (M70/74)

Synonyms:	Gasser revolver M1870/74 or M1874
Adoption date:	February 1875(?)

Data from *Waffen-Instruktion für die Artillerie und die Train-Truppen des k.k. Heeres* (Vienna, 1882)

Length:	320mm (12.6in)
Weight:	1,350g (47.6oz) empty
Barrel length:	184.3mm (7.26in)
Chambering:	11 × 36mm, rimmed
Rifling type:	six-groove, concentric, RH
Depth of grooves:	0.18mm (0.007in)
Width of grooves:	3.85mm (0.152in)
Pitch of rifling:	one turn in 415mm (16.33in), RH
Magazine type:	rotating cylinder
Magazine capacity:	6 rounds
Loading system:	single rounds
Front sight:	open barleycorn
Back sight:	fixed notch
Muzzle velocity:	205m/sec (673ft/sec)
Bullet weight:	20.3g (313gn)

was octagonal, and the backsight was moved to the rear of the frame immediately ahead of the hammer.

11mm Armeerevolver Modell *1870/74 (M70/74)*

This was the finalized form of the revolver designed by Leopold Gasser (1836–71), and accepted for service in the Austro-Hungarian Army on 14 August 1870 to arm NCOs and cavalrymen 'without carbines', the mounted artillery NCOs, *Militärfuhrwesenkorps* personnel, and trumpeters of the *Jäger-Bataillone*.

The double-action lock mechanism was adapted from the French Lefaucheux pinfires, and the barrel was held to the open-top frame by a combination of a Colt-like transverse wedge through the cylinder axis pin and a short screw running longitudinally into the frame-tip beneath the cylinder. This method was comparatively weak, and there is evidence to show that the use of the 1867-pattern carbine cartridge – even in a special reduced-charge form – strained the gun unduly.

Guns made in 1870–74 had wrought iron frames and chambered the standard 11mm cartridge, which consisted of a 20.3g lead bullet and a straight brass case with a prominent rim. A charge of 1.5g. of black powder was used; loaded rounds measured 46.6mm overall and weighed about 28g.

Revolvers made in the early 1870s were 375mm long, had 235mm barrels and weighed about 1,520g. However, the frames were subsequently made of steel instead of iron (advancing the designation to M1870/74), and the barrels were eventually shortened by about 50mm. Other identifying characteristics included a lanyard ring attached to the round butt, and wood grips retained by a single bolt. The backsight was an open 'V'-notch block dovetailed laterally into the barrel-top immediately ahead of the cylinder.

An improved cartridge was introduced specifically for revolvers in 1882. A standard 1867-type bullet was inserted in a case measuring a mere 29mm, giving loaded rounds an overall length of 38mm and a weight of 29.4g. The base and rim were strengthened to improve extraction, but the muzzle velocity was reduced to 165m/sec by a reduction in the size of the powder charge. A change may also have been made to the rifling to suit the lower velocity of the M1882 cartridge, as guns are often reported with six grooves – 0.27mm deep and 2.18mm wide – making a turn in only 275mm.

Markings generally include 'L. GASSER', 'WIEN', 'PATENT' and 'OTTAKRING', the last being the site of the factory. Unit marks may be struck letter by letter into the side of the frame. Military-pattern guns were sold to officers in 'Fine Pattern' forms, distinguished by a better finish; these lacked ordnance marks and often bore nothing but Gasser's trademark of an arrow piercing an apple.

Total production of M1870 and M1870/74 revolvers, official issue and privately purchased

alike, is said to have exceeded 100,000 by 1884 and may have approached 200,000 when the M98 Rast & Gasser pattern was eventually substituted. Many of the old 11mm guns survived to serve home-defence units during World War One, and were still to be seen in the Balkans in 1939.

RIFLES

Front-Rank Designs

8mm Repetier-Gewehr Modell *1895 (M95)*
Trials in 1892 failed to convince the Austro-Hungarian authorities that calibres as small as 5.5mm were worthy enough to challenge the established

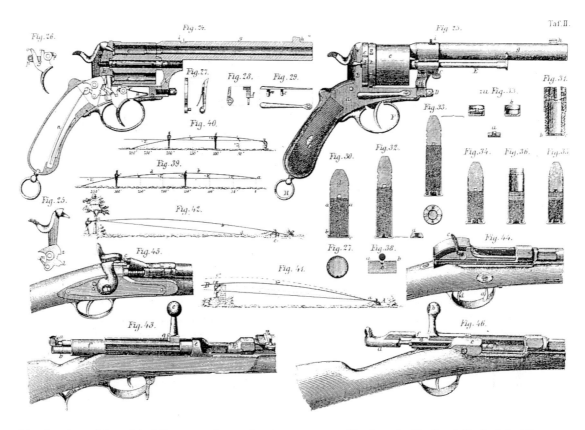

Line drawings of the original Gasser revolver and contemporaneous rifles from the Handbuch für die kais. kön. Artillerie, *Vienna, 1873.*

The M95 Mannlicher rifle. This particular example lacks the wooden handguard that ran from the band to the nose-cap above the barrel. Ian Hogg

8mm pattern. However, it was equally clear that the wedge-lock Mannlichers could not withstand the pressures generated by high-velocity cartridges.

8mm *Repetier-Gewehr Modell* 1895 (M95)

Synonym:	Mannlicher 'straight-pull' infantry rifle M1895
Adoption date:	22 November 1896
Length:	1,273mm (50.12in) *with bayonet* 1,512mm (59.53in)
Weight:	*without sling* 3.78kg (8.33lb) *with bayonet* 4.08kg (8.99lb)
Barrel length:	765mm (30.12in)
Chambering:	8 × 50mm, rimmed
Rifling type:	four-groove, concentric
Depth of grooves:	0.2mm (0.008in)
Width of grooves:	3.5mm (0.138in)
Pitch of rifling:	one turn in 250mm (9.84in), RH
Magazine type:	single-row protruding box
Magazine capacity:	5 rounds
Loading system:	single-sided clip
Cut-off system:	none
Front sight:	open barleycorn
Backsight:	leaf-and-slider type
Backsight setting:	*minimum* 300 *schritt* (225m, 245yd) *maximum* 2,600 *schritt* (1,950m, 2,130yd)
Velocity:	620m/sec (2,035ft/sec) at 12.5m (41ft)
Bullet weight:	15.8g (244gn)

The successful introduction of the M90 carbine, with its short action and dual locking lugs, clearly showed the way forward. In the mid-1890s, therefore, the *Militär-Technische-Komitee* successfully experimented with rifles embodying the 1890-type action, and the M95 was adopted by the *k.u.k Armée* in 1896.

The M95 was made by Österreichische Waffen-fabriks-Gesellschaft and Fémaru Fegyver és Gép-gyár, the guns being marked 'STEYR M. 95' and 'BUDAPEST M. 95' respectively. Several million were made from 1896 to 1918, though large-scale issues did not begin until 1898.

The standard bayonet had a ring-type cross-guard and a 25cm knife blade with the cutting edge on the upper side. NCOs' issue had a small, recurved quillon, extending from the lower part of the guard, and a swivel on the pommel to accept the distinctive cloth knot worn as a badge of rank; bayonets issued to the rank-and-file had plain pommels and short cross-guards. Scabbards were black- or grey-painted steel.

The standard cartridge – the M93 – had a bottle-necked rimmed brass case measuring 50.2mm overall, about 2.1mm shorter than the preceding M90 version. It was loaded with about 2.75g of *Scheibchenpulver* M92 (flake nitro-cellulose pro-pellant) and a jacketed 15.8g lead-core bullet, 31.5mm long. Each cartridge was 76.3mm long and weighed about 29.5g; five were packed in a *Magazin* (clip) held at an angle so that the rims did not interlock. Short serrations on the clip, which could be felt in the dark, showed which side had to be loaded uppermost. The sheet-steel clips weighed 20.2g empty, making them the heaviest of all the clips and chargers that were still in service in 1914.

8mm *Repetier-Gewehr Modell* 1895 (M95)

Internal Arrangements

The rifle has a slender tapering barrel that is screwed into a sturdy receiver containing the bolt mechanism. The bolt consists of a cylindrical body – bored out from front and rear – with full-length ribs on each side. The bolt accepts the bolt head, which carries the twin locking lugs and has a small-diameter rearward extension or 'tail' cut with cam tracks. Lugs on the inside of the body engage the cam tracks to rotate the bolt head lugs during the bolt stroke.

The striker, screwed to the cocking-piece, runs through the centre of the bolt head; it is powered by a coil-spring held in the bolt head by a threaded plug. Specially designed to prevent double-loading, the extractor is a spring-steel claw set into the right-hand rib of the bolt. A safety latch in the left rear side of the bolt body can lock the cocking-piece and the bolt. If the firing mechanism is cocked, the latch also eases the cocking-piece backward until it disengages the sear. The two-part sear and the ejector are formed as a single sub-assembly, driven by a small coil-spring.

The trigger is a slender cranked lever with two lugs projecting upward to prevent the bolt being withdrawn accidentally. The trigger is not attached directly to the body, but is instead held in place by the sear. The magazine case is formed as part of the trigger-guard bow, held to the receiver by two bolts and a small locating stud. An articulated spring-loaded follower is anchored in the base of the magazine, with a clip-retaining catch placed in the rear wall so that its serrated tip projects inside the trigger-guard.

External Appearance

The M95 has a prominent magazine case and a horizontal bolt-handle. It also has a sharp-pointed pistol-grip and a hand guard running the length of the barrel as far as the backsight, where it is held by a small metal plate. A spur on the cocking-piece allows re-cocking in the event of a misfire. The bayonet lug lies on the underside of the nose-band, which also carries a short stacking rod on the left. One swivel lies under the butt; the other appears under the band encircling the mid-point of the fore-end. A collar around the muzzle carries the open front sight, and the back-sight leaf pivots on a sleeve pinned to the barrel. A small battle sight for 500 *schritt* is visible when the leaf is down.

Operation

After the gun is fired, the bolt-handle can be drawn straight back. Although the bolt body is prevented from turning by the side-ribs sliding in their grooves in the receiver, the bolt head is free to turn as the lugs inside the bolt body follow cam tracks in the bolt head tail. This revolves the locking lugs 90 degrees to the left until they align with the side-ribs to allow the bolt and bolt head to be retracted together. The locking lugs are cut on a screw pitch to facilitate extraction in the initial stages of bolt-head rotation, and the initial backward movement of the bolt also compresses the striker spring.

As the bolt opens, the spent case is withdrawn from the chamber and kicked diagonally up and out of the gun by the ejector blade, the right wall of the receiver being cut away appropriately. Backward movement of the bolt is stopped by lugs on the trigger projecting into the bolt way. Thrusting the bolt forward again strips a new cartridge into the chamber from the magazine. The cocking-piece is held back on the sear, completing compression of the striker spring, and the bolt head lugs rotate back into their seats behind the chamber at the end of the stroke. A safety lug on the underside of the bolt ensures that the sear cannot release the cocking-piece until the bolt is fully closed.

When the last round has been chambered, the empty clip falls downwards out of the magazine; however, when the last round has been fired and ejected, the bolt will still close on an empty chamber. Simply opening the bolt and pressing the catch in the front of the trigger-guard bow can eject partially expended clips.

Two clips were loaded in a *Karton* M88 (cardboard packet), labelled with basic manufacturing information. A typical example reads '10 St. 8*mm* M. 93 schf. Patronen' in two lines above 'MF. Wdf. III 1916' and 'B', showing that the '*scharfe Patronen*' or ball ammunition was loaded in March 1916 by the Wolersdorf munitions factory – MF. Wdf., *Munitionsfabrik Wolersdorf* – with propellant supplied by Sprengstoffwerk Blumau ('B'). One hundred and thirty-five *Kartonen* M88 were packed in

The open action of the M95 rifle, showing the locking lugs on the bolt head. Ian Hogg

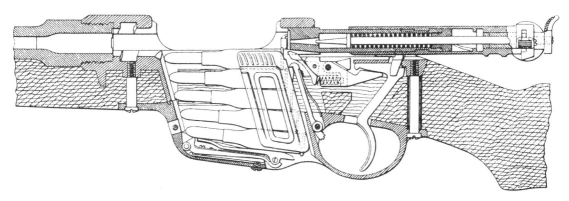

A longitudinal section of the 1895-type action. Konrad von Kromar

Gewehrpatronenverschlag M88 (1888-pattern rifle-cartridge crates).

Rifle grenades could be launched from a cup-type discharger, using a special blank cartridge and a clinometer sight attached to the muzzle. Few details of the Austro-Hungarian grenades are available, though photographs suggest that they were very similar to the German design of 1913 (*see* 'Gewehr 98' entry).

A sniper-rifle adaptation of the M95 was also made in small numbers from 1914 onward. The regulation pattern had a 4× telescope sight made by Reichert of Vienna, held in two ring mounts, but other sights were pressed into service during World War One. The straight-pull action enabled the sight to be mounted lower than those attached to turn-bolt weapons.

The straight-pull action of the M95 rifle was acceptable enough in peacetime, and even under the temporary stress of battle. However, prolonged service showed that it was prone to jamming in mud or snow, and the slender barrel sometimes distorted when it became hot during particularly rapid fire. The Austro-Hungarian authorities would have adopted an improved Mannlicher-Schönauer had not the Italians entered the war in 1915, but the M95 soldiered on into the post-World War One era in the hands of the Austrian, Bulgarian, Czecho-slovakian, Greek, Hungarian and Yugoslav armies.

8mm Repetier-Karabiner Modell *1895 (M95)*
The M95 carbine was essentially a shortened version of the M95 rifle, with the same stock and hand guard design. It had a short barrel, a small

Men of an infantry regiment of the Austro-Hungarian Army pose with M95 rifles. The photograph was taken in Leitmeritz an der Elbe (now in the Czech Republic) some time prior to 1918.

8mm *Repetier-Karabiner Modell* 1895 (M95)

Synonym:	Mannlicher 'straight-pull' carbine M1895
Adoption date:	1897(?)
Length:	1,005mm (39.57in)
Weight:	3.1kg (6.83lb) without sling
Barrel length:	500mm (19.69in)
Chambering:	8 × 50mm, rimmed
Rifling type:	four-groove, concentric
Depth of grooves:	0.2mm (0.008in)
Width of grooves:	3.5mm (0.138in)
Pitch of rifling:	one turn in 250mm (9.84in), RH
Magazine type:	single-row protruding box
Magazine capacity:	5 rounds
Loading system:	single-sided clip
Cut-off system:	none
Front sight:	open barleycorn
Backsight:	leaf-and-slider type
Backsight setting:	*minimum* 300 *schritt* (225m, 245yd) *maximum* 2,400 *schritt* (1,800m, 1,970yd)
Velocity:	580m/sec (1,905ft/sec) at 12.5m (41ft)
Bullet weight:	15.8g (244gr)

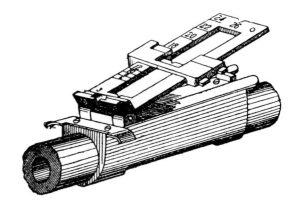

The M95 rifle backsight.

backsight and a plain noseband. A hooked spur on the cocking-piece distinguished 1895-pattern guns from the otherwise similar M90. Swivels were attached to the left side of the barrel band and the left side of the butt-wrist.

M95 carbines made after *c.*1913 apparently conformed to a modified or universal design, with the nose-cap, bayonet lug and stacking rod of the M95 *Stutzen* (short rifle). These accepted the bayonets with auxiliary sights on top of the muzzle ring.

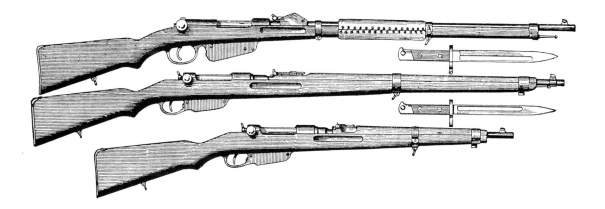

(Above) *Comparative drawings of the M88/90 rifle, M95 rifle and M95 short rifle.* Drawings by J. E. Coombs from the Bannerman catalogue, 1927

(Left) *Men of a mountain unit pose for the camera during World War One. They are carrying M95 short rifles.*

8mm Repetier-Stutzen Modell *1895 (M95)*

The *Stutzen* and *Extra-Corps-Gewehr* – otherwise similar to the cavalry carbine, but slightly heavier – had swivels on the underside of the butt and barrel band. It accepted a special 1895-type knife bayonet with an auxiliary front sight' on top of the muzzle ring to compensate for changes in point-of-impact if the gun was fired with the bayonet attached. Some guns were given additional swivels in 1907, allowing them to be used by mounted or dismounted units alike.

8mm Repetier-Carabiner Modell *1890 (M90)*

The adoption of the M1888 infantry rifle turned thoughts toward a carbine. The experimental guns of 1889 were simply cut-down rifles, but the weakness of their 'wedge' or dropping-bar lock ensured that few were made. An answer was found in a new action submitted to the *Militär-Technische-Komitee* in April 1889 by Ferdinand von Mannlicher. Although the 8mm rimmed cartridge and the clip-loading feature were retained, the locking system relied on two vertically-locking

---

8mm *Repetier-Carabiner Modell* 1890 (M90)

Synonym:	Mannlicher 'straight-pull' cavalry carbine M1890
Adoption date:	23 December 1890
Length:	1,005mm (39.57in)
Weight:	3.29kg (7.25lb) empty
Barrel length:	498mm (19.61in)
Chambering:	8 × 52mm, rimmed
Rifling type:	four-groove, concentric
Depth of grooves:	0.2mm (0.008in)
Width of grooves:	3.5mm (0.138in)
Pitch of rifling:	one turn in 250mm (9.84in), RH
Magazine type:	single-row protruding box
Magazine capacity:	5 rounds
Loading system:	single-sided clip
Cut-off system:	none
Front sight:	open barleycorn
Backsight:	leaf-and-slider type
Backsight setting:	*minimum* 300 *schritt* (225m, 245yd) *maximum* 2,400 *schritt* (1,800m, 1,970yd)
Velocity:	580m/sec (1,900ft/sec) at 12.5m (41ft)
Bullet weight:	15.8g (244gn)

8mm *Repetier-Carabiner Modell* 1890 (M90)

Internal Arrangements

The 1889-patent locking system consists of a bolt which can be rotated inside a sleeve as the handle is pulled back. The turning motion is controlled by lugs inside the sleeve and cam-ways cut in the rear of the bolt body. Two substantial lugs lock in the receiver ring directly behind the barrel.

External Appearance

The carbine has a distinctive one-piece stock with a shallow pistol grip that comes to a characteristic point. The bolt-handle projects horizontally above the trigger-guard and the magazine is made integrally with the trigger-guard. The cylindrical cocking-piece has a small knurled rim, and a safety lever head projects from the left side of the bolt-handle base.

A short finger groove is cut into the fore-end, and a simple nose-cap is fitted. The 1890-pattern carbine – unlike the otherwise similar short rifles – lacks a hand guard, and the swivels are held laterally by studs running through the fore-end and pistol-grip. The backsight is a quadrant pattern.

---

lugs on the bolt. This was far sturdier than the bar lock and could handle the then-experimental 8 × 52mm M90 smokeless-powder cartridges – each loaded with 2.75g of *Zylinderpulver* M90 – without difficulty.

Field trials were very successful. The few minor teething troubles were overcome, and the M90 carbine was ordered into series production. Many thousands were made, survivors being relegated to the *Landwehr* and *Honvéd* after the introduction of the perfected 1895-pattern carbines. A few M90 guns were apparently converted in the early 1900s to fire the M93 cartridge, when the later leaf-type backsight was fitted. In addition, a special knife-bladed socket bayonet was made in small numbers during World War One.

8mm Repetier-Stutzen Modell *1890 (M90)*

This was issued to the *Kaiserliche Marine* some time prior to September 1894, when the relevant shooting instructions were published, and may explain why the existence of an 'M93 torpedo-boat rifle' (usually identified as a Kropatschek) is so often claimed. The M90 short rifle was essentially similar to the 1890-pattern carbine described previously, but had a screwed barrel band held by a spring and a pin. The nose-cap – longer than the carbine type – was retained by a transverse bolt and had a bayonet lug on the right side; a ball-tipped stacking rod lay beneath the muzzle, and swivels lay beneath the butt and a band. The standard bayonet was a variant of the M1888 with a muzzle ring diameter of only 15.2mm.

**8mm *Repetier-Stutzen Modell* 1890
(M90)**

Data: similar to *Repetier-Karabiner* M. 90, (qv).

A drawing of the M1890 carbine. Konrad von Kromar

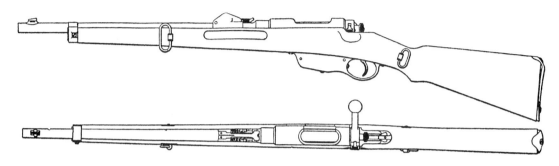

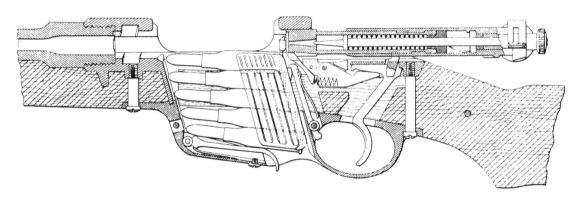

A longitudinal section of the M90. Konrad von Kromar

8mm Extra-Corps-Gewehr Modell *1890* (M90)

Adopted on 14 December 1891 for the *k.u.k. Gendarmerie* and the *Militär-Polizei-Wachkorps*, this was a minor adaptation of the M90 short rifle without the stacking rod on the nose-cap. Though it shared the modified M1888 bayonet, with a small-diameter muzzle ring, it also had a trap in the

butt with a laterally pivoting gate. Guns of this pattern survived in the hands of the gendarmerie until the end of World War One, supplemented by 6.5mm Mo. 91 Mannlicher-Carcano cavalry carbines captured from the Italians.

Second-Rank Patterns

8mm Repetier-Gewehr Modell *1888/90* (M88/90)

No sooner had mass-production of the M1886 begun than the results of the small-calibre trials

**8mm *Extra-Corps-Gewehr Modell* 1890
(M90)**

Data: similar to *Repetier-Karabiner* M90, (qv).

8mm *Repetier-Gewehr Modell* 1888/90 (M88/90)

Synonym:	Mannlicher 'straight-pull' infantry rifle Model 1888/90
Adoption date:	December 1890(?)
Length:	1,280mm (50.39in) *with bayonet* 1,510mm (59.45in)
Weight:	*without sling* 4.35kg (9.59lb) *with bayonet* 4.72kg (10.41lb)
Barrel length:	765mm (30.12in)
Chambering:	8 × 52mm, rimmed
Rifling type:	four-groove, concentric
Depth of grooves:	0.2mm (0.008in)
Width of grooves:	3.5mm (0.138in)
Pitch of rifling:	one turn in 250mm (9.84in), RH
Magazine type:	single-row protruding box
Magazine capacity:	5 rounds
Loading system:	single-sided clip
Cut-off system:	none
Front sight:	open barleycorn
Backsight:	leaf-and-slider type
Backsight setting:	*minimum* 300 *schritt* (225m, 245yd) *maximum* 1,800 *schritt* (1,350m, 1,475yd)
Velocity:	615m/sec (2,015ft/sec) at 12.5m (41ft)
Bullet weight:	15.8g (244gn)

drew attention to the grave mistake in accepting the obsolescent 11 × 58mm pattern.

Work ceased in 1887 to allow tests to begin with a variety of rifles, including Schulhof and Mannlicher-Schönauer patterns. However, the trials board recommended the simplest solution – a new small-calibre version of the 1886-type Mannlicher service rifle – and the M1888 rifle was duly adopted.

The original M1888 was chambered for a cartridge loaded with black powder, and had sights graduated to 1,700 *schritt*; the auxiliary long-range sight was marked for distances of 1,800–2,500 *schritt*. The adoption of smokeless powder in 1890 forced changes to be made in the sights, and most existing guns were modified in 1891–92 by attaching plates marked for the M90 cartridge over the old 1888-type black-powder graduations on the side of the quadrant-base. This created the M88/90 rifle.

Guns made after 1891 ('M90') had newly made sights graduated to 1,800 *schritt*. Alterations were also made to the safety system and applied retrospectively to many of the conversions when they returned for repair.

The 1888-type rifles, original or converted for the M90 cartridge, were issued with a knife bayonet with a 25cm blade. Some bayonets had pommel loops and extended quillons for NCOs' knots; others were plain short-guard examples for the

A longitudinal section of the M1888 rifle.
Konrad von Kromar

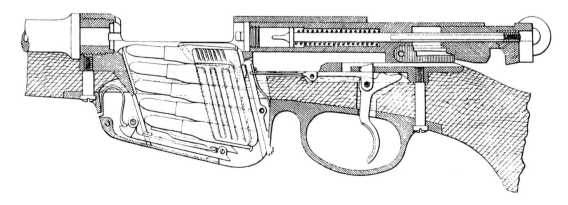

8mm *Repetier-Gewehr Modell* 1888/90 (M88/90)

Internal Arrangements

The barrel screws into the receiver, which accepts a straight-pull bolt. The bolt body is hollowed to receive the striker and its spring. A locking bar or strut is pivoted on the underside of the bolt, and a long spring-steel extractor is retained in a slot on the right side.

The grip piece carries the stubby bolt-handle. A cylindrical extension runs forward from the grip-piece body into the bolt, but is prevented from turning by a flanged actuator that enters a dovetailed recess in the locking bar. The striker is anchored in the cocking-piece, which slides between two horns beneath the rear of the grip piece.

The bolt-retaining catch and the safety lever both lie on the rear left side of the receiver. The safety locks the bolt if the action is at rest, or, alternatively, prevents the cocked striker being released by the sear. A special intercepting notch was added to the safety in 1890 to prevent accidents if the trigger was pressed with the safety latch only partly engaged; original guns will fire the moment the safety is released, but modified examples cannot.

The trigger is a simple double-pressure type. The articulated cartridge follower pivots in the front of the magazine well, and the head of the clip-retaining catch protrudes at the rear. Accurate positioning of the receiver and base plate is assured by two bolts, and by a locating tongue on the receiver which forms the upper front surface of the magazine.

External Appearance

The guns have one-piece stocks, with a finger groove in the fore-end, but lack barrel guards. The gap between the magazine and the large trigger-guard is a good identifying feature. Two screwed bands are used, and the nose-cap has a bayonet lug on the left side. A short stacking rod protrudes beneath the muzzle. Swivels lie beneath the butt and the central band.

The quadrant backsight is set for 300 *schritt* when the leaf is in its lowest position. An unmarked 500-*schritt* setting is accompanied by graduations from 600–1,800 *schritt* on the left edge of the leaf, for use with the open front sight. Graduations from 2,000–3,000 *schritt* on the right of the leaf serve in conjunction with a slider extension and a conical auxiliary long-range sight on the right side of the central barrel band.

Operation

Drawing back the bolt-handle pulls back the guide piece, lifting the locking bar out of engagement in the bolt-way floor before allowing the guide piece and the bolt to slide backward. The extractor pulls the spent case clear of the breech until it strikes a shoulder on the left side of the receiver and is thrown up and to the right – no separate mechanical ejector is necessary. Compression of the striker spring occurs largely as the guide piece moves back in relation to the bolt during the unlocking motion.

The bolt is returned to strip a new round from the magazine into the chamber and then stops against the breech face. However, the guide piece still has sufficient travel to re-engage the locking bar and complete striker spring compression as the cocking-piece is held back on the sear. Pressing the trigger then fires the gun, assuming that the safety latch has not been applied.

The cartridge clip drops downwards out of the magazine after the last round has been chambered, but can be expelled at any time – together with unfired cartridges – simply by opening the action and pressing the clip-release catch head on the back surface of the magazine.

rank and file. A three-piece cleaning rod was carried separately in the soldier's pack, and a felt-lined canvas guard was often laced around the fore-end to protect the hand from barrel-heat and interference with the sight picture.

The original Mannlicher action was simple and easy to make, but was too weak to handle ammunition loaded with smokeless propellant and proved to extract badly. Consequently, it was replaced rapidly by guns with rotary locking systems.

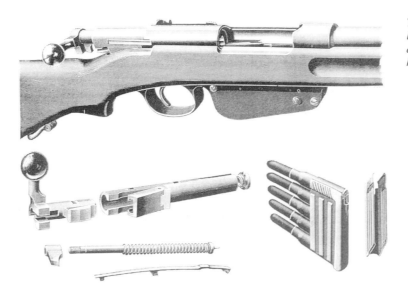

The M1888/90 rifle showing the locking wedge beneath the bolt directly ahead of the horizontal handle. Engineering

8mm Repetier-Gewehr Modell *1886/90* (M86/90)

Originally adopted on 20 June 1886, the M1886 Mannlicher infantry rifle was the first service pattern to embody the wedge- or bar-lock (*see* '8mm *Repetier-Gewehr Modell* 1888/90 [M88/90]' entry). The rifles were about 1,320mm long and weighed about 4.55kg without the short-bladed bayonet.

External appearance was essentially similar to the later M1888, though the original sights were graduated only to 1,500 *schritt* for the 1877-type 11mm cartridge. The long-range auxiliary sights could be used for ranges of 1,600–2,300 *schritt*.

The introduction of the M1886 proved to be a disaster, owing to the appearance of the French Mle 86 (Lebel) rifle only a few months after the Austro-

Men of Infanterie-Regiment Nr 49 pose with musical instruments and six stacked M86/90 Mannlicher rifles.

A longitudinal section of the 1886-type Mannlicher rifle. Konrad von Kromar

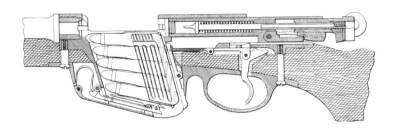

Hungarians had finished development work. Although the Mannlicher was a much better combat weapon than the Lebel, it had been designed around an antiquated cartridge and proved impossible to improve satisfactorily. Production continued only until the advent of the improved 8mm rifle in 1888, but would not have been undertaken at all had not the need to re-arm front-line units with magazine rifles been so pressing.

Most of the 93,000 11mm-calibre M1886 rifles were hastily rebuilt to M1888 standard, creating the M86/88, and then altered for smokeless ammunition after *c.*1892 to create the M86/90. Owing to the reduced diameter of the cartridge base, the magazine case of the conversions is noticeably shallower than the original pattern. Few, if any, magazine rifles chambering the 11 × 58mm round were still available in 1914.

Obsolete Patterns

Magazine Rifles
Sizeable quantities of 1881-type experimental Kropatschek infantry rifles were issued for field trials with the Austro-Hungarian Army, but few (if any) survived into World War One. Field trials guns were about 1,290mm long, had 793mm

barrels and weighed about 4.5kg empty. The magazines held eight rounds and the ramp-and-leaf sights were graduated to 1,600 *schritt*. A 'Navy Rifle M1884' and a 'Torpedo-Boots Gewehr', said to have been issued from 28 October 1893 to the crews of torpedo-boats, are often identified as Kropatscheks. The *Kaiserliche Marine* was still comparatively small in 1890, and the reissue or modification of field trials rifles may have been enough to satisfy the initial demand for magazine rifles. Details are still lacking.

11mm Gendarmerie-Repetier-Karabiner Modell *1881*
Four thousand of these carbines were ordered for the *königlich Ungarnische Landesvertheidigung* or gendarmerie, the order being given to Österreichische Waffenfabriks-Gesellschaft on 19 June 1881. Issue was subsequently extended to gendarmerie in Bosnia-Herzegovina, and then to the *k.k.* (Austrian) *Gendarmerie* on 17 March 1882.

The Kropatschek action was based on the 1871-pattern Mauser rifle, relying on the abutment of the bolt-guide rib on the receiver bridge to lock the mechanism when the bolt handle was turned down. It had a tube magazine beneath the barrel and a pivoting cartridge elevator in the breech.

11mm *Gendarmerie-Repetier-Karabiner* Modell **1881**

Synonyms:	Kropatschek gendarmerie carbine, M1874/81 or M1881
Adoption date:	19 June 1881
Length:	1,044mm (41.1in)
	with bayonet
	1,518mm (59.76in)
Weight:	3.4kg (7.5lb) empty
	with bayonet
	not known
Barrel length:	560mm (22.05in)
Chambering:	11 × 36mm, rimmed
Rifling type:	six-groove, concentric
Depth of grooves:	0.18mm (0.007in)
Width of grooves:	3.7mm (0.146)
Pitch of rifling:	one turn in 525mm (20.65in), RH
Magazine type:	tube beneath barrel
Magazine capacity:	6 rounds
Loading system:	single rounds
Cut-off system:	none
Front sight:	open barleycorn
Backsight:	ramp and leaf-and-slider
Backsight setting:	*minimum*
	200 *schritt* (150m, 165yd)
	maximum
	1,600 *schritt* (1,220m, 1,310yd)
Muzzle velocity:	305m/sec (1,000ft/sec)
Bullet weight:	24g (370gn)

A smaller version of the rifles being tested by the Austro-Hungarian Army in the same period, but with a single barrel band and the bolt handle turned downward, the Kropatschek gendarmerie carbine resembled the earlier Fruwirth. However, the striker had an integral cocking-piece instead of being struck by an external hammer. The action was also appreciably stronger than that of the Fruwirth, chambering the 11mm bottlenecked M1877 carbine cartridge instead of the weaker straight-case M1867. By 1900, surviving guns had been replaced by 1890-pattern Mannlichers discarded by the Army.

Single-Shot Guns

11mm Infanterie- und Jäger-Gewehr Modell *1877 (M77)*

Trials of improved cartridges in 1875–77 led to changes in the rifle, though virtually the only external change made was the new backsight. However, rifles chambering the new long-case M1877 (11 × 58mm) cartridge would not fire the original short-case M1867 (11 × 42mm) pattern, though

11mm *Infanterie- und Jäger-Gewehr* Modell **1877 (M77)**

Synonym:	Werndl M1877 infantry rifle

Data for an 1873-model rifle from Rudolf Schmidt, *Die Handfeuerwaffen, II* (Basle, 1878)

Adoption date:	1878
Length:	1,264mm (49.76in)
	with bayonet
	1,740mm (68.5in)
Weight:	*without sling*
	4.2kg (9.26lb)
	with bayonet
	4.7kg (10.36lb)
Barrel length:	845mm (33.27in)
Chambering:	11 × 58mm, rimmed
Rifling type:	six-groove, concentric
Depth of grooves:	0.2mm (0.008in)
Width of grooves:	3.7mm (0.146in)
Pitch of rifling:	one turn in 724mm (28.5in), RH
Magazine type:	none
Loading system:	single rounds
Front sight:	open barleycorn
Backsight:	leaf-and-slider with ramped base
Backsight setting:	*minimum*
	200 *schritt* (150m, 165yd)
	maximum
	2,100 *schritt* (1,575m, 1,720yd)
Muzzle velocity:	440m/sec (1,445ft/sec) at 12.5m (41ft)
Bullet weight:	24g (370gn)

unconverted guns were deliberately left in the *Landwehr* and *Honvéd* to expend existing supplies. The improved cartridge was issued from 25 December 1878, but 1877-pattern rifles were not delivered in quantity until after this change had been made.

The rifles were issued with the standard 1873-type sabre bayonets, which had recurved 465mm blades and adjuster-screw cross-guards with flattened quillons. A loop in the pommel signifies NCOs' issue, accepting the strap of the bayonet knot; loops were customarily attached by bolts,

although 'field promotion' examples were made by bending wire into a hole drilled through the pommel. Many M1867 and M1870 bayonets were shortened when old Werndl rifles were altered for the M1877 cartridge in the 1880s, but can be recognized by fullers that run almost to the blade-tip.

The standard 1877-pattern cartridge had a 58mm bottlenecked brass case with a prominent rim. It measured 78mm overall and weighed about 42.5g with its 24g round-nose lead bullet and a charge of 5g of black powder. The adoption of improved

---

11mm *Infanterie- und Jäger-Gewehr Modell* 1877 (M77)

Internal Arrangements

One of the few rotary-breech designs ever to reach service, the Werndl relies on a massive drum turning on an axis pin inserted horizontally from the rear of the box-like receiver. The drum-face is cut with a slight bias, enabling it to move back slightly from the breech during the opening stroke. A hammer mounted inside the lock plate strikes a spring-loaded firing pin running through the drum, which is held in the open or closed positions by an internal detent.

A groove in the surface of the drum gives access to the chamber when the breech is open. The opening movement also operates the extractor, which is one of the worst features of the design. An extractor blade is attached to a rod, running laterally beneath the chamber, which is in turn fixed at right angles to a short arm ending in a stud. The stud can slide freely along a groove cut transversely in the surface of the breech drum. When the drum is opened, however, the end of the groove forces the stud downward to rotate the extractor blade and push a spent case clear of the chamber.

External Appearance

The 1877-pattern rifle, identical externally to the 1873 pattern except for the large sliding-leaf backsight, is easily distinguished by the drum breech. The barrel is held in the one-piece stock with two bands and a nose-cap. A bayonet lug lies on the right side of the muzzle, and there are swivels beneath the middle band and butt. The hammer is mounted inside the back-action lock plate, though it is still offset to the right side of the centreline. Guns destined for the *Jäger* have a spur on the trigger-guard extending rearward; infantry weapons have plain rounded guards.

Operation

Starting with the gun in the fired state, the hammer is retracted to half cock. The breech-drum handle on the left side of the gun is rotated a third of a turn to the right, extracting the cartridge from the chamber as it nears the end of its opening movement. The motion is awkward, but becomes easier with practice; however, satisfactory extraction can only be achieved by 'snapping' the block open with as rapid a movement as possible. The spent case is extracted manually, simply by canting the gun. A new round is pushed down the inclined feed way into the chamber and the drum is rotated back to the left. The hammer can be retracted to full cock and the gun is ready to fire again.

The M1873 Werndl Rifle, with a Jägerspur *on the trigger guard.* Hans-Bert Lockhoven

propellant in 1881 is said to have raised velocity slightly (apparently to about 465m/sec) and also to have allowed the backsight graduations to be increased to 2,200 *schritt*. However, very few guns of this type have been reported.

Although huge quantities of Werndl rifles and carbines were still available in 1914, the strength of the breech system – even the perfected 1873 design – was not sufficient to allow 8 × 50mm conversions to be made.

The distinctive backsights of the 1867 (1), 1873 (2) and 1877-pattern rifles (3).

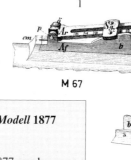

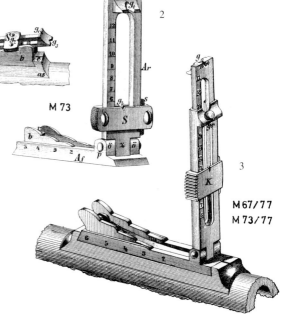

11mm *Kavallerie-Karabiner Modell* 1877 (M77)

Synonym:	Werndl M1877 cavalry carbine
Adoption date:	1878
Length:	990mm (38.98in) *with bayonet* 1,465mm (57.68in)
Weight:	*without sling* 3.185kg (7.02lb) *with bayonet* 3.375kg (7.44lb)
Barrel length:	570mm (22.44in)
Chambering:	11 × 36mm, rimmed
Rifling type:	six-groove, concentric
Depth of grooves:	0.2mm (0.008in)
Width of grooves:	3.7mm (0.146in)
Pitch of rifling:	one turn in 525mm (20.67in), RH
Magazine type:	none
Loading system:	single rounds
Front sight:	open barleycorn
Backsight:	leaf-and-slider with ramped base
Backsight setting:	*minimum* 200 *schritt* (150m, 165yd) *maximum* 1,600 *schritt* (1,200m, 1,310yd)
Muzzle velocity:	305m/sec (1,000ft/sec)
Bullet weight:	24g (370gn)

11mm Kavallerie-Karabiner Modell *1877 (M77)*

This was approved to replace the 1867 and 1873-pattern guns. The new bottlenecked cartridge was interchangeable with its straight-case predecessor, but was loaded with the standard M1877 bullet and was much more powerful. Consequently, the 1877-pattern Kropatschek-designed sight was graduated to 1,600 *schritt* compared with only 600 *schritt* for the original carbine round.

11mm Infanterie- und Jäger-Gewehr Modell *1873/77 (M73/77)*

The drum breech of the M1867 rifle (qv) proved to be very susceptible to fouling, becoming increasingly difficult to rotate until it jammed altogether.

In addition, constructional weaknesses were discovered in the receiver and backsight leaves snapped too often.

The action was extensively revised in 1872–73 by Antonín Špitálsky, head of the technical section of the Steyr factory. The sight base and the sight leaf were strengthened; the receiver sides were flattened; the lock plate was redesigned with an internal hammer; and the bayonet was improved by using an internal coil-spring in the press-stud instead of an external 'L' type.

The new rifle was formally approved on 10 February 1873. It was similar to its predecessor, though the location of the hammer and modifications to the receiver casing were obvious differences.

Troubles encountered with the original Wildburger-pattern rifle cartridge, in addition to jams and extraction failures, were solved only when a strengthened Roth-type case appeared in 1874. However, the introduction of the improved M1877 cartridge in December 1878 led to the re-chambering of many 1873-type rifles – although no external alterations were necessary, except for substituting a 2,100-*schritt* backsight. Converted M1873/77 and new M1877 rifles are difficult to distinguish unless the date of manufacture is evident.

11mm Kavallerie-Karabiner Modell *1873/77* (M73/77)

Originally adopted on 6 November 1874, the revised or 1873-pattern carbine had a low-profile hammer and a nose-cap. Owing to the absence of barrel bands, one swivel was anchored through the fore-end and the other lay on the trigger-guard. Most surviving 1873-type carbines were altered to chamber the bottlenecked cartridge adopted in December 1878, acquiring sights graduated to 1,600 *schritt*. Known as the M1873/77, these could also handle the improved M1878/81 (or M1881) round adopted on 7 July 1882. However, as the two cartridges were interchangeable, no changes were

---

11mm *Infanterie- und Jäger-Gewehr Modell* 1873/77 (M73/77)

Data: *see* 11mm *Infanterie- und Jager-Gewehr Modell* 1877 (M77) entry.

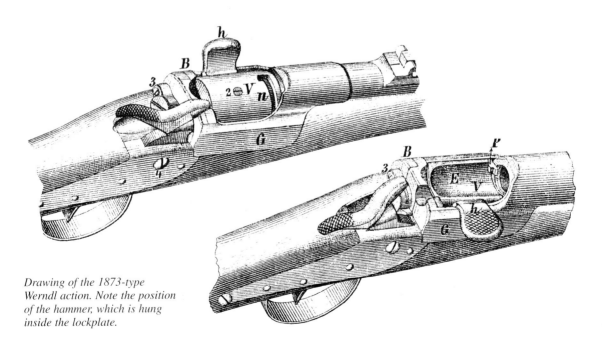

Drawing of the 1873-type Werndl action. Note the position of the hammer, which is hung inside the lockplate.

**11mm *Kavallerie-Karabiner Modell*
1873/77 (M73/77)**

Data: *see* 11mm *Kavallerie-Karabiner Modell*
1877 (M77) entry.

*A typically cheery propaganda postcard produced in
the early days of World War One:* Für Heimat und
Reich *('for Homeland and Empire'). The man on the
right holds an 1867-type Werndl rifle, with a spurred
trigger-guard and an M73 backsight.*

needed in the sights. Some M1873/77 carbines
subsequently received new nose-caps, with a lug
on the right side for the M1873 (common) or
M1886 (rare) sabre bayonet.

11mm Extra-Corps-Gewehr Modell *1873/77
(M73/77)*

Originally adopted in November 1874, the gendarmerie rifle had a small nose-cap retained by a
spring, a swivel beneath the butt and a sling-loop

attached to the fore-end. The trigger-guard had a
spur and a socket bayonet – the *Stichbajonett für
Extra-Corps-Gewehr* M1873 – could be locked
around the front sight. The guns were 1,005mm
long (1,480mm with the bayonet fixed), had
566mm barrels, and weighed 3.25kg. A cleaning
rod was carried beneath the muzzle.

**11mm *Extra-Corps-Gewehr Modell*
1873/77 (M73/77)**

Data: generally similar to the 11mm *Kavallerie-
Karabiner Modell* 1873/77 (M1873/77) (qv).

11mm Infanterie- und Jägergewehr Modell *1867/77 (M67/77)*

The Werndl-Holub breech system was officially
adopted on 28 July 1867, a decision which was
widely acclaimed in Austria. The subsequent manufacture of guns in the new Steyr factory contributed greatly to its prosperity.

The principal feature of the M1867 was its
drum-type breech-block, which, while sturdy and
secure, compromised extraction (*see 'Infanterie-
und Jagergewehr Modell* 1877 (M77)' entry).
Unlike the perfected 1873-type, which relied on an
internal detent, the drum of the 1867-type Werndls
was held in the open or closed positions by a long
leaf spring acting on flats on the head of the
breech-drum axis pin. The spring ran down the
upper tang behind the receiver.

The 1867-pattern Werndl rifle had a straight-
wrist one-piece stock, and a back-action lock with
an external hammer. Infantry guns had a plain
rounded trigger-guard, whereas *Jäger* issues had a
long spur extending backward from the guard.
There were two screwed barrel bands and a nose-
cap; one swivel lay under the middle band; another swivel appeared beneath the butt; and a cleaning
rod protruded at the muzzle.

An attachment lug on the right side of the muzzle accepted the M1867 sabre bayonet, which had
a recurved 58cm blade and a cross-guard with a
flattened quillon. The improved M1870 pattern

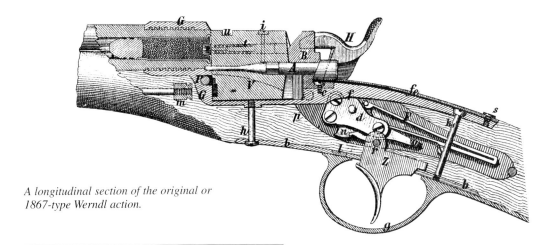

A longitudinal section of the original or 1867-type Werndl action.

11mm *Infanterie- und Jägergewehr Modell* 1867/77 (M67/77)

Synonym:	Werndl M1867/77 infantry rifle
Adoption date:	1878
Length:	1,280mm (50.39in)
	with bayonet
	1,855mm (73.03in)
Weight:	*without sling*
	4.375kg (9.65lb)
	with bayonet
	5.1kg (11.24lb)
Barrel length:	845mm (33.27in)
Chambering:	11 × 42mm, rimmed
Rifling type:	six-groove, concentric
Depth of grooves:	0.18mm (0.007in)
Width of grooves:	3.7mm (0.146in)
Pitch of rifling:	one turn in 724mm (28.5in)
Magazine type:	none
Loading system:	single rounds
Front sight:	open barleycorn
Backsight:	leaf-and-slider with ramped base
Backsight setting:	*minimum* 200 *schritt* (150m, 165yd) *maximum* 2,100 *schritt* (1,575m, 1,720yd)
Velocity:	440m/sec (1,444ft/sec) at 12.5m (41ft)
Bullet weight:	24g (370gn)

had a lighter blade and an adjusting screw – instead of a ball finial – above the muzzle ring. Both patterns will be found with the pommel loops that characterize their use by NCOs.

The 11 × 42mm M1867 cartridge was 60.6mm overall and weighed 32.4g when loaded with a 20.3g bullet and 4g of black powder. The original backsight leaf was graduated to 1,200 *schritt*, with a 'V'-notch in the top edge for 1,400 *schritt*. Chronograph figures taken by the Austrians in 1873 gave an average muzzle velocity of about 450m/sec. However, most surviving rifles, known as 'M1867/77', were adapted in the early 1880s to chamber a new long-body 11 × 58mm cartridge adopted on 25 December 1878.

11mm Karabiner Modell *1867/77 (M67/77)*
Originally adopted at the same time as the infantry rifle, this had a nose-cap but lacked barrel bands. A knob appeared on the hammer instead of a spur. Werndl carbines originally chambered a short-case necked 11mm cartridge loaded with a 20.3g bullet and 2.1g of black powder. This developed appreciably less power than the 11 × 42mm M1867 rifle pattern. The adoption of a more powerful cartridge, in December 1878, eventually led to a change in the chambering and a new backsight instead of the original 600-*schritt* type. Modified carbines were usually known by the designation 'M1867/77'.

11mm *Karabiner Modell* 1867/77 (M67/77)

Synonym:	Werndl M1867/77 carbine
Adoption date:	1878
Length:	991mm (39in)
	with bayonet
	1,465mm (57.68in)
Weight:	*without sling*
	3.185kg (7.02lb)
	with bayonet
	3.73kg (8.22lb)
Barrel length:	566.5mm (22.3in)
Chambering:	11 × 36mm, rimmed
Rifling type:	six-groove, concentric
Depth of grooves:	0.18mm (0.007in)
Width of grooves:	3.7mm (0.146in)
Pitch of rifling:	one turn in 526mm
	(20.7in), RH
Magazine type:	none
Loading system:	single rounds
Front sight:	open barleycorn
Backsight:	leaf-and-slider with
	ramped base
Backsight setting:	*minimum*
	200 *schritt* (150m, 165yd)
	maximum
	1,600 *schritt*
	(1,200m, 1,310yd)
Muzzle velocity:	300m/sec (985ft/sec)
Bullet weight:	24g (370gn)

11mm Extra-Corps-Gewehr Modell *1867/77 (M67/77)*

Chambered successively for the M1867, M1877 and M1881 carbine cartridges, this gun was originally adopted in 1867. Made by Ferdinand Fruwirth in Vienna and Josef Werndl in Steyr, it lacked a barrel band and had a small nose-cap. A swivel lay beneath the butt, but the front sling-anchor point was a loop held in the fore-end by a lateral bolt. The earliest guns had spurred trigger-guards, but not the later versions. The M1854 Lorenz or M1873 '*Extra-Corps*' socket bayonets could be locked around the front sight. The guns were 987mm long, or about 1,457mm with the bayonet attached; they had 566mm barrels and weighed about 3.2kg.

MACHINE-GUNS

Front-Rank Designs

8mm Maschinengewehr Modell *1907/12 (M7/12)*

Originally patented in Austria–Hungary in 1900, in the handgun form, the Schwarzlose machine-gun was successfully tested by the *Militär-Technische-Komitee* in 1905 and issued for field trials in 1905–06 before being adopted in 1907. The design was good enough to serve in Sweden and the Netherlands prior to World War One, being made under licence in both countries. Österreichische Waffen-fabriks-Gesellschaft also supplied 6.5mm-calibre guns to Greece during the Balkan Wars.

The first version, the M7, was characterized by a 'notched' or cutaway receiver and also incorporated a mechanical oil pump to lubricate the cartridges just before they entered the chamber. Changes were made to the feed system, and the revised M7/12, accepted in 1912, had a straight-top receiver and a simplified oil-pad system instead of the pump mechanism.

The M7 and M7/12 machine-guns were often issued with a shield, hinged beneath the gun, which protected the gunners at the expense of considerable additional weight – made of 7mm steel plate, the shield weighed about 40kg. The legs of the tripod could be released to allow the mount to lie directly on the ground, the centreline of the bore being a mere 28cm high.

The Schwarzlose was a remarkably successful machine-gun considering the absence of a breech-lock. This was partly due to the short barrel and the comparatively weak Austro-Hungarian 8mm cartridge, and was also bought at the expense of lubricating the ammunition. Maximum effective range was regarded as about 2,500m, which was substantially less than the 3,500–4,000m claimed for the Maxim, Vickers and comparable weapons.

However, the Schwarzlose machine-guns attained a reputation for durability and were still serving several armies when World War Two began. Most

The M7/12 Schwarzlose machine-gun. Ian Hogg

8mm *Maschinengewehr Modell* 1907/12 (M7/12)

Synonyms:	Schwarzlose machine-gun M7/12 or M1907/12
Adoption date:	1912
Length:	1,070mm (42.13in) without flash hider
Weight:	19.3kg (42.55lb) without mount
Barrel length:	528mm (20.79in)
Chambering:	8 × 50mm, rimmed
Rifling type:	four-groove, concentric
Depth of grooves:	0.2mm (0.008in)
Width of grooves:	3.5mm (0.138in)
Pitch of rifling:	one turn in 250mm (9.84in), RH
Loading system:	fabric belt.
Belt capacity:	100 or 250 rounds
Front sight:	open barleycorn
Backsight:	tangent-leaf type
Backsight setting:	*minimum* 400 *schritt* (300m, 320yd) *maximum* 2,400 *schritt* (1,800m, 1,970yd)
Muzzle velocity:	575m/sec (1,885ft/sec)
Bullet weight:	15.8g (244gn)
Cyclic rate:	400rds/min

8mm *Maschinengewehr Modell* 1907/12 (M7/12)

Internal Arrangements

Only a massive return spring and a system of folding links delay opening of the Schwarzlose breech. The breech-block, which runs back against the spring, is attached to a fixed transverse bar in the receiver by two links. When the gun fires and the breech-block begins to move backward, the rear or 'Y'-link (attached to the block behind the transverse bar) lifts the front link. The two links are joined well ahead of the fixed bar and the unfolding motion is consequently at a mechanical disadvantage; this slows the opening movement of the breech sufficiently to allow the chamber pressure to drop to a safe level, assisted by a short barrel. The breech-block extracts the spent case as it runs back, ejecting it downwards to the left.

The initial movement of the 'Y'-link cams the striker post backward, withdrawing the striker-tip from contact with the primer of the chambered round. A lug on the underside of the breech-block withdraws a new cartridge from the feed belt as the breech opens. The cartridge is rammed forward into the chamber on the return stroke and the links revolve back to their closed position. If the trigger is being held back, the gun fires again; alternatively, if the trigger has been released, the sear bar intercepts the striker bar to hold the striker back against its spring.

External Appearance

Among the most obvious features of the M7/12 Schwarzlose are the high round-backed breech cover (which can be lifted upward after a radial catch has been released) and the faceted extension to the receiver that carries the folding handgrips. A push-button trigger is mounted on the back plate. The charging handle lies on the right side of the receiver above the feed-spool, which has a sturdy folding cover. A finger wheel controls the elevation of the backsight leaf by turning a small pinion, which engages a toothed sector plate on the rear right side of the sight base; the front sight is simply a blade on top of the short water jacket. Most guns originally had a long conical flash hider.

The M7/12 Schwarzlose is mounted on a compact tripod with locking bars to prevent the unit collapsing during use. A wheel acting on a toothed quadrant controls elevation, while traverse is simply a matter of moving the rear of the gun laterally across a sector plate.

of these machine-guns had been successfully converted for much more powerful ammunition.

Second-Rank Designs

Maxim-Maschinengewehr Modell *1904 (M04)*
The Maxim was the earliest successful automatic machine-gun – so successful, indeed, that it was still serving in the 1960s in modified forms (for example, the Vickers Gun). The operating system

was a brilliant conception; though bulkier than many later designs, the additional space around the working parts allowed air to circulate and debris to accumulate without jamming the mechanism.

An 8×50mm gun of this type was demonstrated in Vienna in July 1888, when 13,500 rounds were fired in an endurance trial with only a few jams. A broken mainspring was replaced after 7,281 rounds and the striker failed at 10,233, but the performance was outstanding – particularly as the secrecy-

---

Maxim-*Maschinengewehr Modell* 1904 (M04)

Internal Arrangements

The barrel is attached to an open rectangular box, which slides longitudinally inside the receiver as the gun fires. The box contains a folding-link 'toggle joint' locking system and a complicated breech-block incorporating a vertically-moving cartridge carrier.

When the Maxim fires, the barrel and the breech box move backwards for about 25mm. An external cam-lever attached to the main action pivot, projecting through the rear right side of the receiver, strikes an abutment to break the toggle joint downward. The barrel and the breech box are then brought to a halt, but the links in the locking system continue to fold, cocking the internal hammer as they do so. The folding of the links also draws the breech-block back from the barrel. As the breech-block moves back, the cartridge carrier, still in its uppermost position, begins to extract the spent case from the chamber and pulls a new cartridge backwards from the belt-feed mechanism above the barrel. At the limit of the backward movement, the breech-block stops and the cartridge carrier, which has followed a cam-plate attached inside the receiver, drops downwards with the assistance of a spring under the receiver cover. Movement of the carrier aligns the new cartridge with the chamber and the extracted case with the ejection tube.

The action is closed by a fusee mounted on the left side of the receiver, consisting of a large coil-spring connected to the main action pivot by a short chain. The breech-block runs up against the barrel, the barrel and the breech box are thrust forward, and the links in the toggle joint are cammed back into their locked position. Bell-crank levers on the side of the breech-block raise the cartridge carrier as the connecting link reaches the end of its travel. Careful attention to design allows the spring catches in the cartridge carrier face to slip over the rims of the cases until the upper pawls can lock over the rim of a new cartridge ready to retract it from the belt. The spent case pushes the previous spent case (which is being held by a leaf spring) forward and out of the ejection tube beneath the rear of the barrel jacket.

External Appearance

The Maxim is the archetypal machine-gun, with a deep squared receiver and a large-diameter barrel jacket containing about 2.4lt of water. A tube-and-slider system in the top of the jacket, running parallel with the barrel, prevents water leaking at excessive angles of elevation or depression whilst simultaneously allowing steam to escape. The belt-feed mechanism lies on the right side of the receiver, customarily enveloped in rounded sheet-steel guides, and the toggle-breaking cam lever (doubling as the charging handle) lies on the rear right side; the fusée assembly on the left side of the receiver is protected by a shallow cover.

Spade grips and the trigger lever are mounted on the back plate. Sights consist of an open barleycorn on the front of the barrel jacket, and a folding bar on top of the receiver with a large finger-wheel adjuster on the notch-block. The Austrian guns were originally mounted on tripods, but were subsequently relegated to improvised pintle mounts in static positions.

The action of the Maxim machine-gun. In the upper drawing, the chamber is empty – though a new cartridge is ready to be withdrawn from the feed belt and a spent case is ready to be expelled. In the second drawing, the cartridge has been withdrawn from the belt and the feed-block has dropped to align it with the chamber. The closing stroke of the action will load the cartridge into the chamber and force one of the spent cases out of the ejection tube.

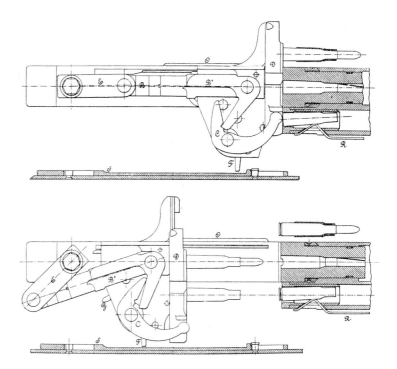

obsessed Austro-Hungarians had declined to provide details of their new 8mm cartridge!

Maxim had had a few thousand cartridges loaded in Britain to enable him to supply a working prototype, but the British ammunition was found to be more powerful than the Austro-Hungarian type once the gun had been delivered to Vienna. Changes hastily made during the early stages of testing miraculously allowed the endurance trial to be undertaken so successfully.

About 160 similar guns were purchased in 1889 from the Maxim-Nordenfelt Guns & Ammunition Co., Ltd, these being chambered for the 8 × 50mm M1888 black-powder cartridge. The guns weighed about 22.5kg and were apparently accompanied by 21.3kg tripods with a central tubular bar extending beyond the gun-mounting bracket. The fabric belts held 334 rounds, subsequently reduced to 250.

The Maxims were converted for M90 smokeless-propellant ammunition in 1891–92, receiving new sights, and were redesignated M89/91. Few additional purchases were made, however, as inter-est was switched first to the Škoda and ultimately to the Schwarzlose. Surviving Maxims were altered again in 1904, becoming the *Maschinengewehr* M4 (or M89/4), but the nature of the alterations remains unclear. The guns had been relegated to fortification duties and the siege trains by 1914.

Škoda-Maschinengewehr Modell *1909* *(M09)*

Much more compact than its predecessors, thanks to the elimination of the pendulum lever, the M09 still retained the projecting mainspring tube that had characterized the M93. The charging handle was moved to the side of the breech, an improved backsight was fitted to the back plate, and changes were made to the cooling system. However, the delayed-blowback Škoda still required cartridges to be lubricated before they entered the breech and a large oil tank was provided on top of the receiver. The cyclic rate of 420rd/min could be varied by altering the length of the striker fall.

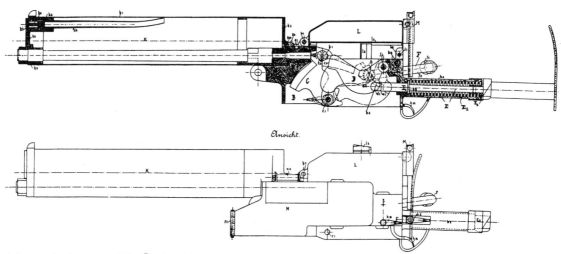

A longitudinal section of the Škoda M09
machine-gun.

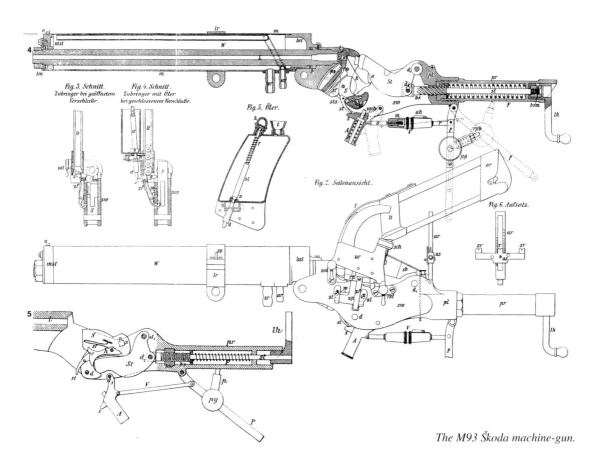

The M93 Škoda machine-gun.

Škoda-*Maschinengewehr Modell* **1909** (M09)

Synonym:	Salvator-Dormus-Škoda machine-gun M1909
Adoption date:	*see* text
Length:	1,045mm (41.15in)
Weight:	15.5kg (34.2lb) without mount
Barrel length:	570mm (22.44in)
Chambering:	8 × 50mm, rimmed
Rifling type:	four-groove, concentric
Depth of grooves:	0.2mm (0.008in)
Width of grooves:	3.5mm (0.138in)
Pitch of rifling:	one turn in 250mm (9.84in), RH
Loading system:	fabric belt
Belt capacity:	100 or 250 rounds
Front sight:	open barleycorn
Backsight:	elevating bar
Backsight setting:	*minimum* 400 *schritt* (300m, 330yd)? *maximum* 2,400 *schritt* (1,800m, 1,970yd)?
Muzzle velocity:	575m/sec (1885ft/sec)
Bullet weight:	15.8g (244gn)
Cyclic rate:	420rd/min (variable)

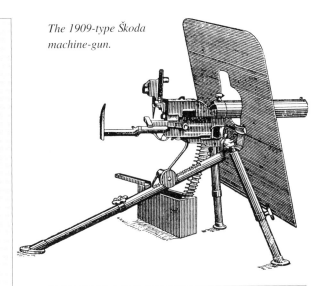

The 1909-type Škoda machine-gun.

Mitrailleuse Patent Škoda *Modell* **1893** (M93)

Synonym:	Salvator-Dormus machine-gun M1893
Adoption date:	15 October 1893
Length:	not known
Weight:	not known
Barrel length:	570mm (22.44in)
Chambering:	8 × 52mm, rimmed
Rifling type:	four-groove, concentric
Depth of grooves:	0.2mm (0.008in)
Width of grooves:	3.5mm (0.138in)
Pitch of rifling:	one turn in 250mm (9.84in), RH
Magazine type:	detachable gravity-feed 'charger'
Magazine capacity:	20 or 30 rounds
Loading system:	single rounds, inserted manually
Front sight:	open barleycorn
Backsight:	folding leaf with a slide
Backsight setting:	*minimum* 300 *schritt* (225m, 245yd) *maximum* 2,000 *schritt* (1,500m, 1,640yd)
Muzzle velocity:	550m/sec (1,805ft/sec)
Bullet weight:	15.8g (244gn)
Cyclic rate:	100–350rd/min (variable)

The tripod was a robust three-position design with integral elevation and traverse adjusters, telescoping legs, and a detachable armoured shield. In its lowest position, the unit virtually lay on the ground. An optical sight made by C.P. Goerz of Wetzlar could be fitted when required.

Obsolete Designs

Mitrailleuse *Patent Škoda* Modell *1893 (M93)*
Patented by Archduke Karl Salvator and Georg, Ritter von Dormus, the Škoda machine-gun was an extraordinary design even by the standards of its day. The frame contained a hybrid two-part breech-block that drew inspiration from both the Remington Rolling Block and the Martini, and the rate of fire could be adjusted manually.

Mitrailleuse **Patent Škoda** *Modell* **1893 (M93)**

Internal Arrangements

When the gun fires, the pressure generated on the base of the cartridge case pushes back against the front part of the breech. The breech-piece can move radially around its pivot, but the motion is opposed by a powerful coil-spring pressing forward on the rear or pivoting part of the block. Pressure of the spring and the disposition of the pivots – plus a certain amount of friction – prevents the two parts of the breech-block disengaging until the chamber pressure has dropped to a safe level.

The front part of the breech-block rotates back beneath the feed way, extracting the spent case and throwing it clear of the breech before returning to push a new round forward from the gravity-feed magazine case or 'charger'. Finally, just before the mechanism comes to rest, the rear part of the breech-block is raised to prop the front portion behind the chamber. A hammer-like lever is then released to propel a short striker into the primer of the chambered round, and the cycle can begin again.

The rate reducer, one of the oddest features of the Škoda machine-gun, consists of a weighted pendulum attached to a swinging link beneath the breech-block by a screw-adjustable tubular union. The pendulum can be adjusted by sliding its weight along the carrying rod; at the lowest point, with the union at its longest, the firer can actually catch the pendulum after each shot. The mechanism was usually set to give a cyclic rate of 100–150rd/min, though as many as 350rd/min can be fired if the pendulum lies in its uppermost position and the union has been shortened to its limits.

External Appearance

The Škoda cannot be mistaken for any other gun, owing to the pendulum mechanism protruding beneath the receiver. The charging handle is mounted on the rear of the mainspring rod, contained in a small-diameter tube extending backwards from the receiver. A skeletal magazine or 'charger' is attached to the left side of the receiver, projecting up and back towards the firer. This can be replenished with single rounds, but not while the gun is being fired.

An auxiliary oil tank and a spring-needle drip system in the top of the receiver lubricates cartridges before they enter the chamber, as the Škoda – like many delayed-blowbacks – extracts poorly in dry or dusty conditions. The backsight has a large leaf, pivoted to the back of the receiver, with extending arms on both sides of the slide. A bracket for the mount is clamped to the water jacket, which is bronze on the earliest guns but sheet-steel on later examples. The jacket is held in place by a large nut at the muzzle, relying on greased leather washers to prevent leakage. The inlet and outlet tubes for the water-circulating system lie between the mounting bracket and the underside of the receiver.

The Škoda could be mounted on a light tripod for field use, on a pivoting pillar mount for use in fortifications, on a wheeled carriage, and in an armoured cupola or casement. Free-mounted guns often had a shoulder stock clamped to the mainspring tube. Guns of this type were still nominally on the inventory of the Austro-Hungarian Army and Navy in 1914, but whether any were used in anger is unclear.

The Škoda was comparatively simple and may have worked reliably in favourable conditions; the two-part breech-block was compact and sturdy, and it is worth remembering that the Madsen machine-gun – which has a similar block-type mechanism – has always been highly regarded for its reliability. Unfortunately, the movement of the pendulum prevented the Škoda machine-gun being mounted close to the ground. The cooling system was another bad feature, as it relied on water being pumped continuously through the barrel jacket with a separate manually operated force-pump.

The official manual suggested that the Škoda had an effective range of 2,250m, but also that firing should be restricted to no more than three minutes at a time – illustrating just how little was known in 1893 about the tactical use of machine-guns!

6 Bulgaria

Without its own manufacturing facilities, Bulgaria was forced to rely on imports. Turkish and Russian influences had gradually given way to Austro-Hungarian and German assistance when the First Balkan War began in 1912.

HANDGUNS

Bulgaria was one of the principal users of the Parabellum pistol, acquiring small quantities in 1901. They had 120mm barrels, grip safeties, and the Bulgarian Arms above the chamber. However, they are believed to have been taken from Deutsche Waffen- und Munitionsfabriken's normal commercial production runs.

Sales figures given by Deutsche Waffen- und Munitionsfabriken in 1911–12 indicated that only 1,300 7.65mm Parabellums had been supplied since 1901. In view of the existence of Bulgarian Old Model Parabellums numbered in the 20200–20700 group, the 7.65mm New Model Bulgarian guns, and the change to 9mm in 1910, it is unlikely that more than fifty Old Models were sup-plied to Bulgaria – though claims as high as 500–1,000 have been made.

The Later Parabellums

The 7.65mm 'New Model' Parabellum was adopted by Bulgaria on 5 October 1908, immediately after independence from Turkey had been declared. Approximately 1,500 New Model 7.65mm pistols, with grip safeties, were delivered for officers' use in 1909. They were numbered from '1' in a separate sequence and bore the distinctive pavilioned Arms above the chamber.

Shortly before the outbreak of the First Balkan War of 1912, when Bulgaria, Serbia, Greece and Montenegro allied against Turkey, the Bulgarian Army ordered 10,000 9mm Parabellums from Deutsche Waffen- und Munitionsfabriken. Numbered 1–5000 and 1C–5000C, the pistols were standard *Pistolen* 08, lacking stock lugs, but had lanyard rings on the left side of the butt. The Deutsche Waffen- und Munitionsfabriken monogram was struck above the chamber and simplified Arms appeared on the front toggle-link.

A top view of the 1906-pattern Bulgarian Parabellum, showing the ornate chamber crest. Dr Rolf Gminder

A 1906-pattern Bulgarian Parabellum. Originally made with a 7.65mm-calibre 120mm barrel, this gun has been altered to 9mm. Dr Rolf Gminder

The 1910-type Bulgarian Parabellum had a lanyard ring on the butt. Dr Rolf Gminder

RIFLES

Most of the early Peabody-Martini and Berdan II rifles, acquired from Turkey and Russia respectively, were out of service by 1914 – though survivors were doubtless pressed into second-line service.

The Mannlichers

Substantial quantities of 1888-pattern Mannlicher rifles were ordered from Österreichische Waffen-fabriks-Gesellschaft in the early 1890s, followed by 160,000 M95 rifles (apparently including carbines and short rifles) in 1897. Supplementary orders are believed to have been placed prior to 1912 and again, to make good losses sustained in the Balkan Wars, in 1914. Many of the older M1888 rifles were converted to handle the M93 smokeless-propellant cartridge in the early 1900s, approximating to the Austro-Hungarian M88/90. The Bulgarian Mannlichers were identical with their Steyr-made prototypes, but had the national Arms over the chamber and the 'crown/F' ciphers of Ferdinand I.

MACHINE-GUNS

Very little is known about the inventory of the Bulgarian Army, except that a solitary Maxim was acquired from the Maxim-Nordenfelt Arms & Ammunition Co., Ltd in *c*.1889. Chambered for the 10.6×58mm rimmed Berdan cartridge, then the Bulgarian service pattern, the Maxim was apparently only intended for trials and nothing else was done until the early 1900s.

The Maxims

About 150 M1907 guns were purchased from Deutsche Waffen- und Munitionsfabriken in 1907–08. Chambered for the 8×50mm rimmed Austro-Hungarian ammunition, these 1901-pattern commercial Maxims were followed in 1910–12 by an unknown number of 8mm M1909 guns – the standard Deutsche Waffen- und Munitionsfabriken commercial derivative of the German MG08. Apart from guns retained in static roles or mounted aboard warships, the Bulgarian Maxims were all issued with tripods.

7   Germany

HANDGUNS

Front-Rank Patterns

9mm Pistole Modell *1904 (P04)*

No formal adoption has yet been traced, though an 8,000-gun contract was delivered to Deutsche Waffen- und Munitionsfabriken on 12 December 1904. On 12 May 1905, the Navy Office noted that 'development of the self-loading pistol ... is to be considered as successfully finalized', and

---

9mm *Pistole Modell* 1904 (P04)

Synonyms:	Marine-Modell M1904, Parabellum M1904, M1904 Navy Luger
Adoption date:	1904
Length:	267mm (10.52in)
Weight:	1,010g (35.6oz) empty
Barrel length:	150mm (5.91in)
Chambering:	9 × 19mm, rimmed
Rifling type:	four, later six-groove, concentric
Depth of grooves:	0.125mm (0.049in)
Width of grooves:	2.75mm (0.108in)
Pitch of rifling:	one turn in 250mm (9.84in), RH
Magazine type:	detachable box in butt
Magazine capacity:	8 rounds
Loading system:	single rounds
Front sight:	open barleycorn
Backsight:	sliding two-position notch
Backsight setting:	100m and 200m
Muzzle velocity:	350m/sec (1,150ft/sec)
Bullet weight:	8g (123gn)

---

also that no large-scale deliveries were expected until March 1906.

Although there is a possibility that 2,000 guns were ordered in the summer of 1904, in advance of the order for the perfected or 1906-type coil spring guns, nothing in the *Kaiserliche Marine* budgets indicates that substantial quantities of pistols were delivered to the navy in 1904–05.

Progress was delayed by improvements, and production was initially very slow. The existence of the first manual was not announced in the *Marine-verordnungsblatt* until 15 February 1906. However, a claim that the *Kaiserliche Marine* acquired at least 1,250 of the original riband-spring Parabellums, made many years ago, still meets with strong opposition if challenged.

The contract had called for the delivery of 8,000 guns by March 1906, but Deutsche Waffen- und Munitionsfabriken was soon asking the *Reichs-Marine-Amt* to relax the schedule and the final deadline was extended to the end of May 1906. Any further delay would 'not be tolerated'. It seems most likely that the new coil-pattern mainspring had taken longer to perfect than had been anticipated.

The *Selbstladepistole* 1904 was officially renamed *Pistole* 1904 in February 1907 and, with effect from 22 June 1912, the operation of the safety lever was altered to prevent it being released unintentionally. Although the change is not detailed in surviving *Kaiserliche Marine* papers, it clearly refers to the introduction of 'down-safe' operation.

The manual safety lever had a projection on its tip, which rotated behind a shoulder on the grip-bar to lock the mechanism in place when the safety was

The 'pre-production' Pistole 04 had a spring-lock set into the right toggle grip. Thompson D. Knox

Men of the Torpedo-Division pose for the camera after completing their basic training in May 1911. Note the Parabellum pistols, holsters and magazine pouches.

applied. However, the lever locked in its upper position. Replacing the gun in its holster sometimes pushed the manual catch downwards, unlocking the grip safety and unintentionally freeing the gun to fire. By mid-1912, therefore, the mechanism had been redesigned to work in reverse, preventing the lever – which locked in its lower position – from being reset as readily as its predecessor had been.

The authorities intended to alter all the pistols so that their safety system, even though an additional grip lever was fitted, resembled the perfected 1908-pattern Army Luger. The work was apparently entrusted to Kiel dockyard. Guns serving with naval personnel stationed in the German colonies seem to have been replaced with modified guns, perhaps when the ships or land-based units returned to Germany.

A few unaltered guns may still have been in service when World War One began. In January 1914, however, having had second thoughts, the Navy Office decided that the grip safety of the *Pistole 1904* was to be eliminated entirely. Until the necessary orders had been issued, the imperial dockyards were expressly forbidden to 'disengage the grip safety'. The existence of this order may suggest that two separate changes were to have been made to the safety system of the *Pistole 1904*: it was to have been 'revised' with effect from June 1912 (changing

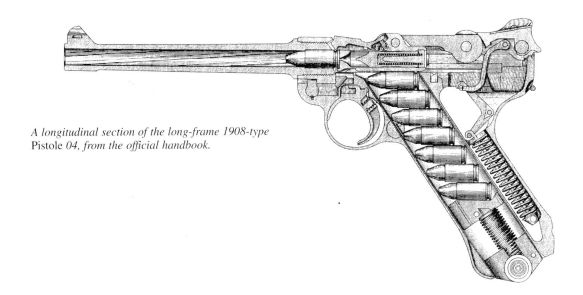

A longitudinal section of the long-frame 1908-type
Pistole *04, from the official handbook.*

it to down-safe operation) and then 'disengaged' altogether, possibly by removing the interlock between the grip- and manual safeties.

9mm *Pistole Modell* 1908 (P08)

Synonyms:	Parabellum M1908, M1908 Army Luger
Adoption date:	22 August 1908
Length:	216mm (8.5in)
Weight:	885g (31.25oz) empty
Barrel length:	100mm (3.94in)
Chambering:	9 × 19mm, rimless
Rifling type:	six-groove, concentric
Depth of grooves:	0.125mm (0.049in)
Width of grooves:	2.75mm (0.108in)
Pitch of rifling:	one turn in 250mm (9.84in), RH
Magazine type:	detachable box in butt
Magazine capacity:	8 rounds
Loading system:	replacement magazine
Front sight:	open barleycorn
Backsight:	fixed open notch
Backsight setting:	50m
Muzzle velocity:	335m/sec (1,100ft/sec)
Bullet weight:	8g (123gn)

9mm Pistole Modell *1908 (P08)*

The German Army adopted the 9mm Parabellum in March 1907 for the four experimental infantry machine-gun detachments, formed in August 1906 and reconstituted as machine-gun companies on 23 May 1907. Each company had seven Maxim machine-guns and ninety-three *Pistolen* 08 in 1910 (later reduced to eighty) and it has been estimated that about 500 pistols were acquired from Deutsche Waffen- und Munitionsfabriken.

Comparative tests undertaken by the *Gewehr-Prüfungs-Kommission* with Frommer and Mauser pistols throughout the summer and autumn of 1907 merely confirmed the superiority of the Borchardt-Luger design. In February 1908, therefore, the '9mm self-loading pistol Luger with flat-nose steel-jacketed bullet' was recommended as a replacement for the *Reichsrevolvers*. The relevant orders were signed by Kaiser Wilhelm II in August 1908.

Deutsche Waffen- und Munitionsfabriken was given an order for 50,000 *Pistolen* 08, 50,000 screwdrivers and 9,000 cleaning rods. At least 3,000 pistols were to be delivered by 31 March 1909, allowing issue to begin once the design of a new holster (*Pistolentasche* 08 or PT08) had been

---

9mm *Pistole Modell* 1908 (P08)

Internal Arrangements

The pistol consists of the barrel and receiver, a frame, and a multi-link locking system attached to the breech-block. When the gun fires, the barrel/receiver and breech-block groups recoil, securely locked together, until the sturdy grips on the toggle strike the ramps formed in the back of the frame. This lifts the transverse joint between the toggle links above the axis of the bore and breaks the lock. The barrel and the receiver are halted by an abutment on the frame, but the toggle links continue to fold upward around a fixed pivot in the rear of the receiver until the breech-block is drawn back from the breech far enough to clear the magazine well. The spent case is ejected upwards.

A coil-spring in the back of the grip, connected by a bell-crank lever and a stirrup to the back toggle link, then propels the breech-block forward to strip a new round into the chamber. The barrel and receiver return to battery, and the lock is reasserted as the toggle links rotate over-centre.

External Appearance

The Pistole 08 is a most distinctive gun, with a slender tapering barrel and the toggle mechanism on top of the action. The chequered walnut grips are attached to a raked extension of the frame, which contains the detachable box magazine. The coil-pattern mainspring and the combined extractor/loaded-chamber indicator of the navy Pistole 1904 are retained, but the grip safety – and the old upward-acting manual lever – has given way to a simple lever acting directly on the exposed laterally-moving sear bar on the left side of the receiver. The sear is easily immobilized by sliding a plate vertically out of the frame-side. The front sight lies on a collar around the muzzle, and a fixed open backsight appears on top of the back toggle link.

Operation

The action functions semi-automatically until the last spent case has been ejected. Apart from the earliest guns, which lacked a mechanical hold-open, the breech stays open after the last shot and the open toggle breaks the firer's sight line.

The magazine-release catch on the left side of the frame behind the trigger is pressed to drop the spent magazine, a new magazine can be inserted in the butt, and the toggles are pulled slightly upwards to disengage the hold-open. The action will then run forward, reloading the chamber as it does so, and the gun is ready to fire. Alternatively, the safety lever set into the rear left side of the frame can be rotated, raising a blade out of the frame side to block the lateraly moving sear.

---

formally approved on 4 April 1909. The last delivery was to be made in March 1911, when a duplicate production line installed in the Erfurt rifle factory would begin to fulfil demand at a rate of 20,000 *Pistolen* 08 annually.

Chamber-dating was introduced in 1910, when the inspectors' marks were transferred to the front right side of the receiver to allow the serial number to take their original place on the left.

About 18,000 *Pistolen* 08 were delivered in 1910, followed by 28,040 in 1911. The quantities delivered in 1909 remain unknown, but sufficient had been delivered by the spring of 1909 to allow issue to begin to the *Maschinengewehr-Abteilungen*

(independent machine-gun detachments) and the machine-gun companies of the infantry regiments. These were followed by the reserve and *Ersatz* (supplementary) machine-gun units, which received their handguns in July 1909.

The infantry, riflemen, pioneers, train and telegraph units received Parabellums in the autumn of 1909 – accounting for all the initial deliveries, so that the cavalrymen had to wait until October 1910. Issues to the cavalry were completed by March 1911, when the foot artillery and *Luftschiffertruppen* (airship detachments) had also received their first guns.

Issue to the field army had been completed by the end of 1911, allowing non-combatants – for

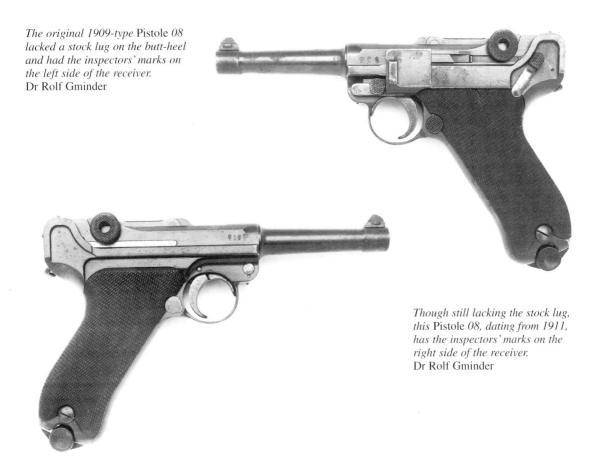

The original 1909-type Pistole 08 *lacked a stock lug on the butt-heel and had the inspectors' marks on the left side of the receiver.*
Dr Rolf Gminder

Though still lacking the stock lug, this Pistole 08, *dating from 1911, has the inspectors' marks on the right side of the receiver.*
Dr Rolf Gminder

example, stretcher-bearers and medical personnel – to receive Parabellums in 1913. By the end of the 1913 fiscal year, therefore, about 153,000 *Pistolen 08* had been delivered from Deutsche Waffen- und Munitionsfabriken and Erfurt.

Officers had always been expected to purchase their handguns privately, and so Deutsche Waffen- und Munitionsfabriken-made Lugers bearing all the characteristics of the Army *Pistole 08* will be found with five-digit numbers and commercial proof marks. Guns purchased privately from Erfurt, however, were numbered in the cyclical military serial number blocks and were indistinguishable from those of NCOs and other ranks. In addition, men who had been commissioned from the ranks were allowed to purchase their original

service pistols if they wished to do so, blurring the distinction between military-issue and privately purchased Parabellums still further.

Field service soon showed the error of omitting the hold-open, which kept the action to the rear when the last round had been fired and ejected. On 6 May 1913, therefore, the *Kriegsministerium* published details of the hold-open 'fitted to *Pistolen 1908* of recent manufacture' and noted that the Erfurt rifle factory had been instructed to transform existing guns when convenient.

Owing to the adoption of the *Lange Pistole 1908*, a stock lug was added on the butt-heels of standard pistols delivered after 4 August 1913. It was also discovered that the height of the front sight and a lack of proper sighting-in before leaving the factory

Pictured 'somewhere in France' in June 1916, these machine-gunners of Bavarian Infanterie-Regiment Nr 25 are armed with Pistolen *08 and bayonets.*

were contributing to poor point-blank shooting. On 12 June 1913, the *Kriegsministerium* decreed that the addition of hold-opens and adjustments to the sights would be undertaken simultaneously.

The sights of guns already fitted with hold-opens were to be adjusted by regimental armourers. Instructions were issued to the Erfurt rifle factory to begin converting guns immediately. The Bavarian Army decided to collect its handguns together by 31 July 1914, ready for wholesale shipment to

Erfurt, but the imminence of war persuaded some units to refuse to surrender their guns unless substitutes could be provided. Consequently, the guns were never shipped to Erfurt and it is tempting to suggest that many of the *Pistolen* 08 that are now encountered without hold-opens were once Bavarian property.

No changes were made to the *Pistole* 08 during World War One, excepting that the sear bar was changed in 1916. The guns performed well enough

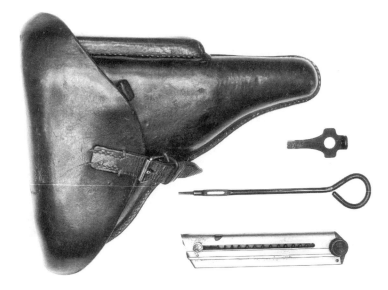

A typical Pistole *08 holster of the World War One period, showing the spare magazine, cleaning rod and combination tool.* Per Jensen

as long as the action was kept reasonably clean, though the exposed sear bar on the left side of the receiver was prone to bind unless lightly coated with oil. However, the Parabellum is needlessly complicated, and the triggers, sears and trigger-plates were often hand-fitted to ensure satisfactory performance.

9mm lange Pistole Modell *1908 (LP08)*

The *Kriegsministerium* had decided as early as 1907 to equip field artillerymen with long-barrel pistols. Once the field army had been issued with *Pistolen 08*, therefore, a team led by Adolf Fischer began development of the 'long Parabellum' and the *Lange Pistole* 1908 was adopted for the armies of Prussia, Saxony and Württemberg in the summer of 1913. The order received imperial assent on 2 July and the Bavarians accepted the gun on 12 September. Each pistol was to be issued with a special combination board-type butt-stock and leather holster.

The long Parabellum was a standard *Pistole* 08 in all respects except for its 200mm barrel and a tangent-leaf backsight, which lay immediately ahead of the receiver. The design of the sight-block required a small step to be milled out of the front upper edge of the receiver, which was subsequently added to all

9mm *lange Pistole Modell* 1908 (LP08)	
Synonyms:	Long Parabellum M1908, M1908 Artillery Luger
Adoption date:	3 June 1913
Length:	317mm (12.48in)
Weight:	1,105g (39oz)
Barrel length:	200mm (7.87in)
Chambering:	9 × 19mm, rimless
Rifling type:	six-groove, concentric
Depth of grooves:	0.125mm (0.049in)
Width of grooves:	2.75mm (0.108in)
Pitch of rifling:	one turn in 250mm (9.84in), RH
Magazine type:	detachable box in butt
Magazine capacity:	8 rounds
Loading system:	replacement magazine
Front sight:	open barleycorn
Backsight:	tangent-leaf type
Backsight setting:	*minimum* 100m *maximum* 800m
Muzzle velocity:	375m/sec (1,230ft/sec)
Bullet weight:	8g (123gn)

The lange Pistole *08, showing its tangent backsight.*

A comparison of the three German Parabellums: (top to bottom) the LP08, P04 and P08. Masami Tokoi

Erfurt-made *Pistolen* 08 (though rarely encountered on Deutsche Waffen- und Munitionsfabriken examples) so that only one type of receiver was needed.

The wartime pistol manual, *Anleitung zur langen Pistole 08 mit ansteckbarem Trommelmagazin (T.M.)* stated that the LP08 'on account of its high

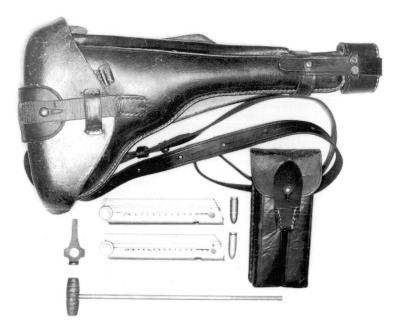

A typical LP08 holster and shoulder stock, complete with magazines, magazine pouch and accessories. This example dates from 1918. Per Jensen

firepower and easy handling' could be used effectively against head-size targets at 600m and that 'accuracy to 800 metres' was possible if the backsight had been adjusted accordingly.

The sights of most of the guns made prior to 1917 could be adjusted with set-screws and capstan tools, facilitating long-range accuracy, but this feature was subsequently abandoned to save production time.

Problems with the sights and hold-opens of the standard pistols delayed work in Erfurt until February 1914, when 209,000 *Lange Pistolen* 1908 were ordered for the field artillery, the airmen and some specialist ancillary units. The first part of the contract – 144,000 guns – was to be completed within five years, 75,000 being made in Erfurt and 69,000 by Deutsche Waffen- und Munitionsfabriken. Another 65,000 guns were then to be made in Erfurt for the *Ersatz* units, the *Landwehr* and the *Landsturm*, but the outbreak of World War One put an end to the scheme almost before it had begun.

Second-Rank Patterns

A few 7.65mm pistols – for example, the Dreyse and the Sauer – were used by some of the state police forces, and the Roth-Sauer saw limited use with the gendarmerie in South West Africa. Apart from the Roth-Sauer, which was a diminutive predecessor of the Austro-Hungarian M7 Roth-Steyr (qv), the guns are included in Part Three of this book on the grounds that they were impressed into military service only during World War One.

7.63mm Mauser-Selbstlade-Pistole C96
Designed by the Feederle brothers in the early 1890s, the Mauser pistol encountered success only after an adaptation of the 7.65mm Borchardt cartridge had been made.

Though tested personally in August 1896 by Kaiser Wilhelm I – a valuable publicity coup – the C96 eventually lost out to the Borchardt-Luger in German service. However, substantial quantities were obtained for field trials: 145 in the summer of 1898 alone, followed by 172 in January 1899 and

---

7.63mm *Mauser-Selbstlade-Pistole* C96

Synonym:	'Broomhandle Mauser' pistol
Adoption date:	none

Data for an 1898 trials gun

Length:	289mm (13.38in)
Weight:	1,005g (35.5oz) empty
Barrel length:	127.5mm (5.02in)
Chambering:	7.63 × 25mm, rimless
Rifling type:	four-groove, concentric
Depth of grooves:	0.14mm (0.055in)
Width of grooves:	3.07mm (0.121in).
Pitch of rifling:	one turn in 250mm (9.84in), RH
Magazine type:	internal box in frame-front
Magazine capacity:	10 rounds
Loading system:	charger or single rounds
Front sight:	open barleycorn
Backsight:	fixed notch
Muzzle velocity:	430m/sec (1,410ft/sec)
Bullet weight:	5.5g (85gn)

Note: most post-1910 guns had six-groove rifling with a depth of 2.85mm (0.112in), making a turn in only 200mm (7.87in).

---

fifty-five 'improved models' in the summer of 1902. Experimental guns were also submitted, including a short-barrel version and a 'lightened pattern for officers' of 1901.

The production pattern had been finalized by the spring of 1897, once the spur hammer had been replaced by a ring pattern and a second lug had been added to the locking block. A plain or 'slab-side' frame was introduced in 1899, followed in 1902 by a revised two-piece firing pin and a safety lever which moved downward to the firing position. A reversion to the panelled frame was made in 1905, followed by a change to six-groove rifling and the adoption, in 1912, of a *neue Sicherung* (NS: new safety) which moved upward to lock, but could only be applied when the hammer was cocked.

The C96 was a clever design with interlocking parts, but was complicated, difficult to machine

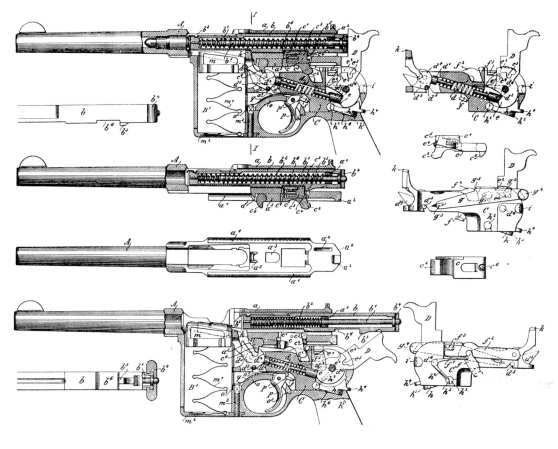

(Above) *A sectional drawing of the Mauser C96 action.*

(Left) *A typical Mauser C96 pistol, showing how the cartridges are fed from the charger into the magazine.*

Opposite page:
(Top) *The Mauser C96 was also tested experimentally as a carbine, though guns of this type ultimately found greater success commercially.*

(Bottom left) *C96 pistol No. 23922, with flat-side frame and holster.* Weller & Dufty Ltd

(Bottom right) *C96 pistol No. 81502, with holster-stock.* Weller & Dufty Ltd

7.63mm *Mauser-Selbstlade-Pistole* C96

Internal Arrangements

When the C96 fires, the barrel and the receiver move back together until lugs on the block locking the bolt in the receiver are cammed down into the frame-floor and out of engagement. The barrel and receiver are halted by an abutment in the frame to allow the bolt to run back alone, ejecting the spent case and rotating the hammer until it can be held on the sear. The backward movement also compresses the return spring, contained in the bolt, against a transverse bar through the receiver.

The spring reasserts itself to reverse the motion of the bolt, thrusting it forward to strip a new round into the chamber, then cams the locking block back into engagement with the bolt as the barrel/receiver group runs back to battery.

External Appearance

The Mauser has a deep slab-sided frame containing a charger-loaded magazine ahead of the small round trigger aperture. The grips are usually finely ribbed, held in the frame by the only bolt in the entire action, and a lanyard ring lies on the base of the butt-strap. Most guns have a tangent-leaf sight on top of the breech, but fixed-sight guns were made in small numbers. The safety lever protrudes alongside the hammer.

Operation

The C96 fires semi-automatically until the magazine has been emptied and the last spent case has been ejected. The bolt is held open mechanically, allowing the magazine to be replenished with single cartridges or a ten-round charger. Pulling the bolt slightly back overcomes the hold-open mechanism and allows the action to close, loading a new round into the chamber and leaving the hammer at full cock. The gun can now be fired or, alternatively, the safety-catch can be applied.

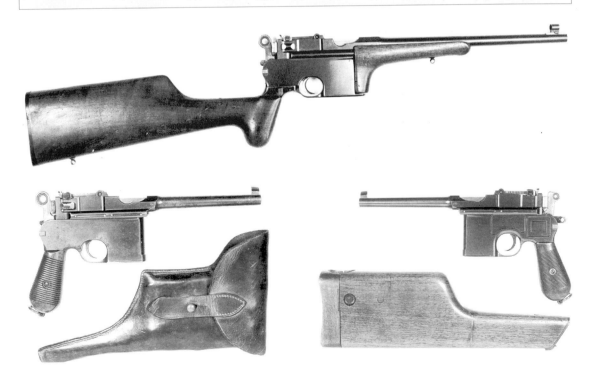

and expensive to make. Many variations in frame panels and hammer design were made in a production life of forty years, and selective-fire versions, developed experimentally during World War One, were offered in quantity in later years.

Magazines holding six, ten or twenty rounds were used. The backsight may have been a fixed open notch, but the tangent-leaf patterns (100–500m or 100–1,000m) were much more popular for use at long range – especially when the gun was accompanied by a holster stock.

Although substantial quantities of the C96 were sold, including 5,000 to the Italian Navy and 1,000

to Turkey, the German authorities remained disinterested until World War One (*see* Part Three: '1915').

Obsolete Patterns

10.6mm Revolver Modell *1879*
Credited to the *Gewehr-Prüfungs-Kommission*, this was a surprisingly basic solid-frame design with a single-action trigger and a radial safety lever on the left side of the frame. No ejection system was fitted, as the firer was required to punch individual spent cases out of the swinging loading

A typical Reichsrevolver *M79, a holster and a cartridge pouch.*
John Walter

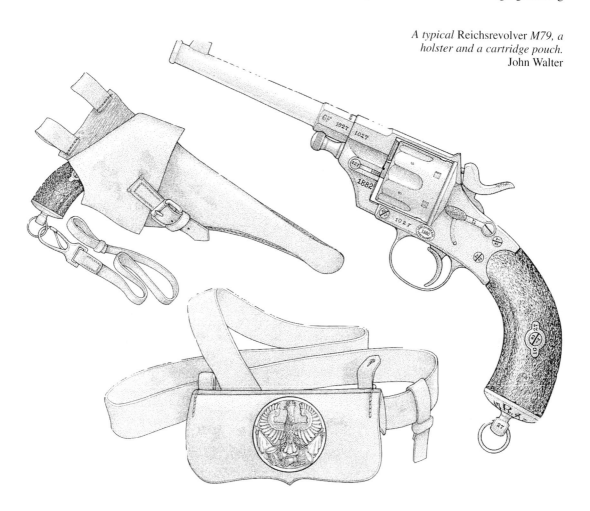

10.6mm Revolver *Modell* 1879

Synonym:	*Reichsrevolver* M1879
Adoption date:	21 March 1879
Length:	338mm (13.31in)
Weight:	1,285g (45.3oz) empty
Barrel length:	180mm (7.09in)
Chambering:	10.6 × 25mm, rimmed
Rifling type:	six-groove, concentric
Depth of grooves:	0.25mm (0.1in)
Width of grooves:	3.6mm (0.142in)
Pitch of rifling:	one turn in 575mm (22.65in), RH
Magazine type:	rotating cylinder
Magazine capacity:	6 rounds
Loading system:	single rounds
Front sight:	open barleycorn
Backsight:	fixed notch
Muzzle velocity:	205m/sec (675ft/sec)
Bullet weight:	17.2g (265gn)

gate on the right side of the frame behind the cylinder. A suitable clearing rod was carried on top of the ammunition pouch, and the half cock notch on the hammer disengaged the locking bolt to allow the cylinder to turn freely.

The M1879 had a lanyard ring attached to the butt-cap and a prominent rib around the muzzle. The gun was intended to arm the cavalry, though issues do not seem to have been made for anything other than field trials until the M1881 holster was approved on 31 August 1881. The dragoons received revolvers immediately, followed by the cuirassiers in February 1885, and the NCOs, standard-bearers and musicians of the infantry regiments some time prior to the advent of *Feldausrüstungsordre* (Field Equipment Orders) of 1885. M1879 revolvers were eventually issued to the field artillery from 24 February 1887.

The M1879 was made by Gebr. Mauser & Cie for the Württemberg Army; and by Schilling, Haenel and Spangenberg & Sauer (together or alone) for the armies of Prussia and Saxony. Examples have also been reported with the marks of J.P.

Sauer & Sohn, the Erfurt rifle factory, and Waffenfabrik Franz von Dreyse. They served until replaced by the *Pistole* 08, though many remained in second-rank service when World War One began. This accounts for the proliferation of ammunition-column and similar unit markings found on back straps or butt-caps.

The standard 10.6mm cartridge, loaded with 1.5g of *neues Gewehrpulver* M1871 until reduced to 1.25g in 1885, had a straight-rimmed centre-fire case loaded with a round-nose lead bullet with two cannelures. The rounds each measured about 37mm overall and weighed about 24g.

10.6mm Revolver Modell 1883

The introduction of this short-barrel derivative of the M1879 remains unclear, and it has even been suggested that no issues were made until 12 March 1891, when the M1883 revolver and the M1891 holster were approved for the dismounted personnel of the field artillery batteries.

The M1883 is mechanically similar to the 1879 pattern, though the barrel was greatly shortened and the frame ahead of the cylinder was cut back;

10.6mm Revolver *Modell* 1883

Synonym:	*Reichsrevolver* M1883
Adoption date:	not known
Length:	272mm (10.71in)
Weight:	925g (32.6oz) empty
Barrel length:	118mm (4.65in)
Chambering:	10.6 × 25mm, rimmed
Rifling type:	six-groove, concentric
Depth of grooves:	0.25mm (0.1in)
Width of grooves:	3.6mm (0.142in)
Pitch of rifling:	one turn in 575mm (22.65in), RH
Magazine type:	rotating cylinder
Magazine capacity:	6 rounds
Loading system:	single rounds
Front sight:	open barleycorn
Backsight:	fixed notch
Muzzle velocity:	187m/sec (615ft/sec)
Bullet weight:	17.2g (265gn)

The Reichsrevolver *M83*. Weller & Dufty Ltd

consequently, the cylinder axis-pin retainer became a vertical leaf spring with a small press-catch. The butt was rounded, though the lanyard ring was usually retained. However, many 'Officer's Model' revolvers of this type were made prior to 1905, with much finer finish, grips of diced rubber or chequered walnut, spring-loaded ejector rods and double-action triggers; Dreyse even made guns with set triggers in spurred guards. Commercial examples of this type do not bear military inspectors' marks.

RIFLES

Front-Rank Patterns

7.9mm Gewehr Modell *1898 (Gew. 98)*

This rifle was developed in the late 1890s to replace the *Gewehr* 88/97. Trials had been undertaken with many small-calibre cartridges, including some as small as 5mm, but all had failed to challenge the 8mm *Patrone* 88.

The finalized *Gewehr* 98 differed greatly from the *Reichsgewehr*, the preceding service rifle. The receiver had a solid bridge; the bolt-handle locked behind the bridge; double locking lugs were forged integrally with the body of the one-piece bolt; a third or 'safety' lug was seated in the receiver ahead

---

7.9mm *Gewehr Modell* 1898 (Gew. 98)

Synonym:	German Mauser infantry rifle M1898
Adoption date:	5 April 1898
Length:	1,255mm (49.41in) *with bayonet* 1,771mm (69.75in)
Weïght:	*empty* 4.085kg (9.01lb) *with bayonet* 4.480kg (9.88lb) with S98
Barrel length:	740mm (29.13in)
Chambering:	7.9 × 57mm, rimless
Rifling type:	four-groove, concentric
Depth of grooves:	0.166mm (0.065in)
Width of grooves:	4.48mm (0.176in)
Pitch of rifling:	one turn in 239mm (9.41in), RH
Magazine type:	internal staggered-row box
Magazine capacity:	5 rounds
Loading system:	single rounds or chargers
Cut-off system:	none
Front sight:	open barleycorn
Backsight:	Lange-pattern tangent-leaf with slider
Backsight setting:	*minimum* 400m *maximum* 2,000m
Muzzle velocity:	879m/sec (2,885ft/sec)
Bullet weight:	9.85g (152gn), 'S' type

Note: weights and dimensions from a specimen tested in Britain in 1908.

---

of the bolt-handle; and a special tangent sight, designed by *Oberst* Wilhelm Lange, replaced the original leaf-and-slider type. The bayonet attachment system was radically altered, the barrel jacket was discarded, and a pistol-grip was added to the butt. Most importantly, a charger was used to load the magazine instead of a clip.

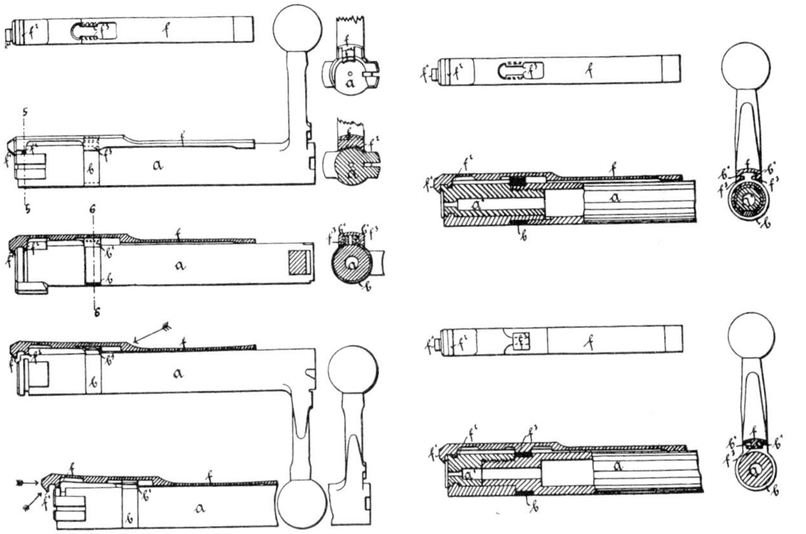

1

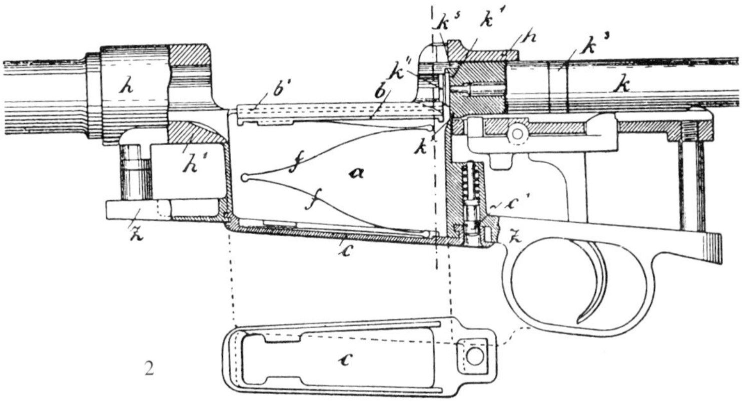

2

Among the features of the Gewehr 98 *were the non-rotating extractor (1), the internal magazine (2), the safety-catch on the cocking-piece (3), and the safety lug on the bolt. These patent drawings dating from 1892–95 capture the essence of the design.* R.H. Korn

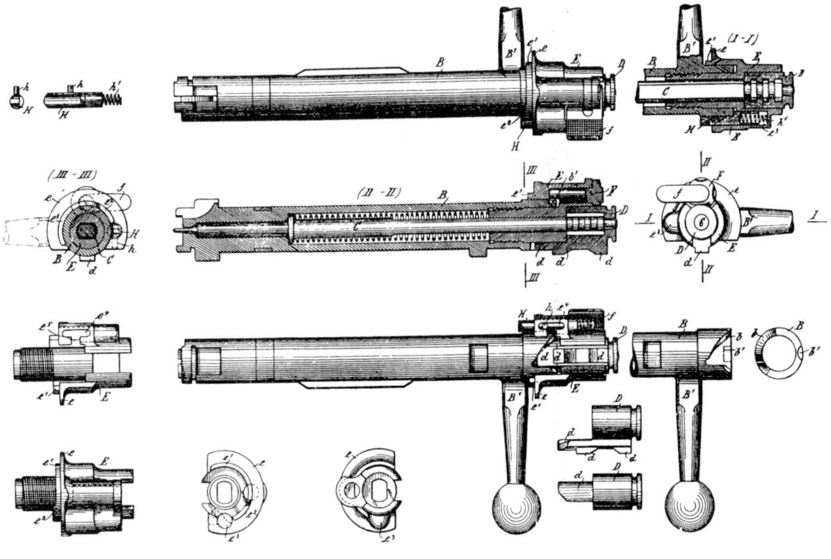

3

7.9mm *Gewehr Modell* 1898 (Gew. 98)

Internal Arrangements

The Gew. 98 receiver is a one-piece forging with the bolt-handle locking down behind the bridge. The barrel screws into the front of the receiver, where its base abuts a specially machined collar that surrounds the entire bolt head except for the extractor passage. The rear edge of the collar is bevelled to assist the loading of a cartridge into the chamber.

The bolt is a solid machined forging, with an integral handle and guide rib. Two lugs on the bolt head lock vertically in the receiver directly behind the breech face. A third or 'safety' lug, on the underside of the bolt body in front of the handle, seats in a well in the receiver-bridge behind the magazine; the third lug makes contact with the back of its recess only if the front lugs shear away.

The lower of the two bolt head lugs and the safety lug are solid, whereas the upper bolt head lug is slotted to allow the ejector to kick a spent case clear of the bolt way. The bolt face is partially rimmed to support the cartridge case, and has a small projection around the ejector-blade slot (diametrically opposite the extractor claw) to ensure that the cartridges do not drop as the bolt is retracted.

The long spring-steel extractor is attached to the bolt body by a collar placed immediately behind the locking lugs. It is prevented from moving longitudinally by a small projecting lip behind the extractor claw engaging a short undercut groove on the bolt body ahead of the locking lugs. The bolt is bored-out from the rear to accept the striker and the mainspring assembly.

The bolt is closed from the rear by the cocking-piece guide or bolt plug, which screws into the bolt body to provide an efficient bearing for mainspring compression. It also carries the 'wing'-type safety mechanism and a large flange to divert gases that may escape from a ruptured primer, out through two large oblong ports in the underside of the bolt body and along the guide-way for the upper (left) locking lug. The cocking-piece is attached to the striker that runs into the bolt body through the cocking-piece guide or bolt plug.

The magazine – a light sheet-steel box – is contained entirely within the stock, the staggered cartridge column being lifted by a light steel follower powered by a flat leaf 'W'-spring mortised into the detachable magazine floor plate.

External Appearance

A long, straight bolt-handle locks down horizontally behind the receiver-bridge. Prominent charger guides are milled in the front of the bridge and a distinctive thumb-clearance cutaway appears in the left receiver wall. A standard Mauser-pattern dismantling catch lies on the rear left side of the receiver body, where it can be pivoted outwards to release the bolt.

The one-piece walnut stock has a pistol-grip and a hand guard running forward from the backsight to slightly ahead of the barrel band. The band is retained by a leaf spring. A marking disc or dismantling washer appears on the side of the butt. Sling swivels lie under the butt and band, and a small hook under the nose-cap can be used to shorten the sling.

The unusual nose-cap and bayonet-bar assembly relies on a long under-muzzle bar that is isolated from the barrel. The bar is attached to an integral nose-piece that is in turn clamped to the stock fore-end by a barrel band retained by a pin and a spring. A half-length cleaning rod, carried beneath the muzzle, can be joined with the rod from another rifle when required.

Operation

Starting with the weapon in its fired state, the bolt-handle is raised to revolve the locking lugs out of their recesses. This motion cams back the cocking-piece, compresses the striker spring and (owing to the design of the cam surfaces on the bolt-handle base and the receiver) begins extraction of the spent cartridge case with considerable mechanical advantage. When the bolt-handle is vertical, the bolt can be drawn back through the receiver-bridge, supported by its guide rib, until progress is stopped as the left (upper) locking lug comes to rest against the bolt-stop. Simultaneously, the ejector contained in the bolt-stop mechanism, working in the locking lug guide-way, passes through the slotted lug and kicks the spent case up and out of the right side of the gun.

The bolt is then returned to strip another cartridge out of the magazine and into the chamber. The cocking-piece is held back on the sear, and a new cartridge in the breech is seated gently by the camming action of the bolt-handle and by the locking lugs entering their seats. The gun is now loaded, cocked and ready to fire. However, if the magazine is empty, the bolt will have closed on an empty chamber owing to the absence of a hold-open. With the bolt open, a charger can be placed in guides milled into the receiver-bridge and five cartridges pushed down into the magazine far enough to be held by the feed lips. The empty charger is automatically thrown clear as the bolt is closed.

The horizontal handle of the *Gewehr* 98 bolt is badly placed for rapid shooting, although even the British *Text Book of Small Arms* (1904) grudgingly admitted that it could be grasped without taking the eyes off the target. However, the position of the handle when fully retracted forces the firer to raise his cheek from the butt – disturbing his aim – and compares unfavourably with the Lee-Enfield.

Large-scale troop trials with 1. *Garde-Regiment zu Fuss*, the *Garde-Jäger-Bataillon*, the *Garde-Schützen-Bataillon* and the *Militar-Scheiss Schule* in Spandau began on 9 February 1899. Plans for large-scale manufacture were made immediately, but though the government enthusiastically voted a large sum to complete rearmament in just five years, the *Kriegsministerium* realized that the hurried introduction of the *Reichsgewehr* had created serious long-term problems. The issue of the new Mausers, therefore, was eventually spread over a decade.

Mass production of the *Gewehr* 98 began in the Prussian government arsenals during 1900. By the end of February 1901, Danzig arsenal was delivering 140 rifles daily, compared with 107 from Spandau and 54 from Erfurt. Issues to the first three army corps were made late in 1901 and the 100,000th government-made *Gewehr* 98 appeared in February 1902.

The new rifle was not adopted in Bavaria until *Prinzregent* Luitpold signed the relevant papers on 2 May 1901. Machinery was installed in the Amberg factory in August and the first guns were delivered to the Ingolstadt fortress in January 1903; the two Bavarian army corps had been entirely rearmed by October 1907.

By 1907, when front-line re-equipment was supposed to have finished in Bavaria, Prussia Saxony and Württemberg, at least one million *Gewehre* 98 had been made. These included the first to be made by Waffenfabrik Mauser AG, which was responsible for 290,000 of the 500,000 guns ordered in 1904. The others were made by Deutsche Waffen- und Munitionsfabriken. Additional orders seem to have been placed in 1910 and 1912, allowing *Gewehre* 98 to reach the Reserve and the *Landwehr*.

Few changes were made to the *Gewehr* 98 or its accessories, although Mauser registered the design of an improved safety feature in May 1901. This allowed lugs on the front portion of the striker head to align with shoulders inside the bolt body when the action was unlocked, preventing the striker head from protruding out of the bolt face, but its precise value has been the subject of differing assessments.

Off to war with the traditional flower posies, these German soldiers carry Gewehre *98 and* S98/05 *bayonets. Two men also wear marksmanship lanyards.*

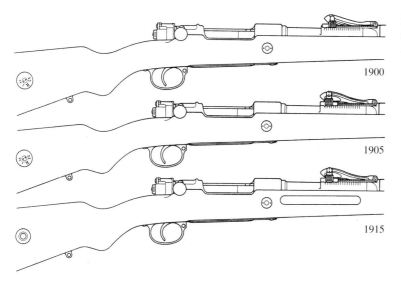

The principal external changes made to the Gewehr *98 prior to 1918.* John Walter

1900

1905

1915

The *Gewehr* 98 initially fired the *Patrone* 88, but the sudden introduction in France of a pointed bullet for the Lebel rifle and the Berthier carbines (the so-called *Balle* D) caused the development of a comparable *Spitzgeschoss* or pointed bullet to begin in Spandau. Streamlining, changes in the propellant and a reduction in bullet weight raised muzzle velocity from 620m/sec to 870m/sec.

Adopted on 25 March 1903, the S-*Patrone* was loaded with 3.135g of nitro-cellulose propellant, with a 9.85g pointed lead-core bullet (jacketed in a cupro-nickel alloy) which was about 28.1mm long. The bottlenecked brass case measured 56.7mm, which gave the complete round an overall length of 80.8mm – fractionally shorter than the combination of the *Patrone* 88 and its round-nose projectile. The loaded 'S' cartridges weighed about 24g, five being carried in a sheet-steel charger weighing 7–7.2g.

Stocks of the 1888-pattern cartridges were gradually expended, as supplies of the new S-*Patrone* were delivered into store. Changeover day was 1 October 1905, although 163 million *Patronen* 88 were still in store a year later.

All service rifles were converted for the new cartridge in 1903–05 in Spandau and Amberg. The backsights were naturally revised at the same time,

as the light, fast-moving bullet of the S-*Patrone* had a flatter trajectory than the heavy, slow-moving bullet of the *Patrone* 88. Service weapons were collected in regimental depots, where their sight leaves were replaced with new components supplied by government workshops. The sight bases of converted guns were graduated from 200m to 2,000m, but the 200m and 300m marks could not be used once new leaves had been fitted. The sights of newly made guns were graduated from 400m upwards.

The 1898-type rifles display a maker's mark and date above the chamber. They were made by the Prussian government arsenals in Danzig, Erfurt and Spandau; by the Bavarian factory in Amberg; by Deutsche Waffen- und Munitionsfabriken AG in Berlin-Charlottenburg and Berlin-Wittenau; and by Waffenfabrik Mauser AG in Oberndorf am Neckar.

Guns may also be found with the markings of C.G. Haenel Waffen- und Fahrradfabrik AG, V.C. Schilling & Cie and Simson & Cie of Suhl (all recruited in 1915); or with the marks of Waffenwerke Oberspree, Kornbusch & Cie of Berlin-Niederschönweide, which was purchased by Deutsche Waffen- und Munitionsfabriken in 1916.

The designation mark 'GEW.98' is struck into the left side of the receiver in Fraktur, ahead of the thumb-clearance cutaway, while *Beschussadler* (proof eagles) will be found on the left side of the barrel and the chamber-area of the receiver.

The first bayonet to be issued with the Gew. 98 was the *Seitengewehr* 98 (S98), adopted in April 1898 in Prussia, Saxony and Württemberg, and in May 1901 in Bavaria. It had a long, slender blade with a pipe-back, and a wood-gripped steel hilt. The cross-guard had a short, back-swept *quillon*, but lacked a muzzle ring.

Machine-gunners were issued with a short knife bayonet known as the *Kurzes Seitengewehr* 98 (KS98), introduced everywhere except Bavaria in March 1901. Pioneers, telegraph troops and field artillerymen had the *Seitengewehr* 98/05 (S98/05) or 'Butcher Knife', issued from November 1905 with a distinctive 38cm swell-point blade. The Reserve had the *Seitengewehr* 84/98 (S84/98), with a 25cm blade and a greatly abbreviated cross-guard. This bayonet seems to have been introduced *c.*1903, and a new version was approved for cavalry use in December 1914.

The *Gewehr* 98 could also fire *Gewehrgranate* (rifle grenades), the first pattern appearing in 1913. This had a *feldgrau*-painted steel body, serrated to fragment on detonation, with the base closed by a brass cup receiving a 45cm copper-washed steel rod. A copper ferrule and a copper plug on the base of the rod expanded into the rifling to provide an efficient gas seal when the special grenade blank was fired.

A tin disc attached to the grenade nose by the igniter plug was used to restrict range to below 200m. The principal safety feature was a small pellet of compressed black powder, which kept the striker away from the detonator until burnt away shortly after firing. However, the element of safety proved to be unreliable – the grenade was apt to ignite if dropped – and an improvement appeared shortly before World War One began.

The 1914-type *Gewehrgranate* had a serrated cast-iron body (45mm diameter) containing the bursting charge, but a percussion-impact fuze was screwed into the nose to give an overall length of 139mm. A range-reducing cupped disc and a 455mm tail rod with a copper gas check were then screwed into the base.

Pictured less than a month before the end of World War One, Heinrich Wagner carries a Gewehr *98 with the muzzle protector in place.*

The grenade was armed on firing by the set-back of a locking ring, which allowed the percussion pellet and the nose of the fuze to overcome a locking ball and move forward from 'safe' to the 'live' position.

7.9mm Radfahrer-Gewehr Modell *1898* (R. Gew. 98)

Made in small numbers for issue to cyclists, this rifle had the bolt-handle turned down against a recess in the stock and sling swivels mounted on the side of the butt and barrel band. It was discontinued upon introduction of the *Karabiner* 98 AZ in January 1908.

Karabiner *98 AZ*

The excessive muzzle blast, flash and recoil encountered in the two short 98-system carbines developed in 1900–04, particularly after the high-velocity S-*Patrone* had been introduced, forced the *Gewehr-Prüfungs-Kommission* to develop an experimental long-barrel gun. Field trials began in June 1906 with 800 carbines, all but 100 being fitted with an *Aufpflanzvorrichtung* (stacking rod).

When the trials ended in the summer of 1907, it was clear that muzzle blast, flash and recoil were more tolerable in the long-barrel weapons than in the short-barrel *Karabiner* 98 A used for comparison. However, the absence of a bayonet attachment was strongly criticized. An improved prototype was successfully tested in the autumn of 1907, becoming the *Karabiner* 98 *mit Aufpflanz-*

7.9mm *Karabiner Modell* 1898 AZ (Kar. 98 AZ)

Synonym:	Mauser M1898 short rifle
Adoption date:	16 January 1908
Length:	1,090mm (42.91in)
	with bayonet
	1,560mm(61.42in) with S98/05
Weight:	*without sling*
	3.725kg (8.21lb)
	with bayonet
	4.15kg (9.15lb) with S98/05
Barrel length:	590mm (23.23in)
Chambering:	7.9 × 57mm, rimless
Rifling type:	four-groove, concentric
Depth of grooves:	0.15mm (0.006in)
Width of grooves:	4.4mm (0.173in)
Pitch of rifling:	one turn in 240mm (9.45in), RH
Magazine type:	internal staggered-row box
Magazine capacity:	5 rounds
Loading system:	single rounds or chargers
Cut-off system:	none
Front sight:	protected barleycorn
Backsight:	tangent-leaf with slider
Backsight setting:	*minimum* 300m
	maximum 2,000m
Muzzle velocity:	about 810m/sec (2,655ft/sec)
Bullet weight:	9.85g (152gn), 'S' type

7.9mm *Karabiner Modell* 1898 AZ (Kar. 98 AZ)

Internal Arrangements

The *Karabiner* 98 AZ is similar mechanically to the *Gewehr* 98, though the external diameter of the receiver ring is only 33mm (1.3in) instead of 35.8mm (1.41in). Sight or touch easily detects this, as the left side of the receiver is straight in the Kar. 98 AZ but has a noticeable 'step' on the Gew. 98 at the juncture of the receiver ring and the receiver side-wall. No important mechanical changes were made prior to 1918, though some guns gained a reciprocating bolt cover.

External Appearance

The *Karabiner* 98 AZ is stocked almost to the muzzle, with a heavy nose-cap, a single spring-retained barrel band, and two transverse recoil bolts – one below the chamber and another behind the barrel band. The bolt-handle knob has been changed from spatulate to a half-sphere, with a chequered underside to improve grip, and a recess in the stock makes the bolt easier to use. A sling aperture is cut through the butt from the left side, a sling ring lies on the left side of the barrel band, and the trigger-guard bow lacks the transverse hole that characterized the Gew. 98.

The one-piece walnut stock has the pistol grip tightened to help the firer control recoil. A wooden hand guard runs the length of the barrel from the receiver ring to a special hinged nose-cap that clamps the 4cm bayonet bar to the stock. A special stacking rod is attached to a metal plate running back from the nose-cap under the fore-end, but no cleaning rod is carried.

A tangent-leaf backsight replaces the *Lange Visier* and a finger groove is cut into the fore-end of carbines made after 1915. The muzzle carries a barleycorn front sight surrounded by a distinctive 'eared' protector, and a small lug retains the *Mündungsschoner* (muzzle protector).

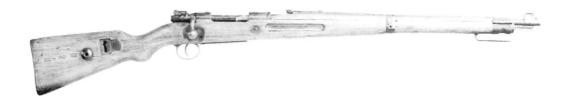

The Karabiner *98 AZ.* Ian Hogg

Taken at the Döberitz firing range in April 1917, this shows men of the Ersatz Bataillon *of* 1. Garde-Fussartillerie-Regiment. *Their* Karabiner *98 AZ are neatly stacked.*

und Zusammensetzvorrichtung ('with bayonet attachment and piling hook') or *Karabiner* 98 AZ. It differed from its predecessors largely in the design of the nose-cap – a special hinged band clamping a standard 4cm bayonet bar, which lay beneath the muzzle but did not contact the barrel directly.

The *Karabiner* 98 AZ replaced the 1902-pattern *Karabiner* 98 A. The first issues were made to the cavalry in the middle of 1909, and the foot artillery received its first guns in the autumn of 1910. When World War One began, *Karabiner* 98 AZ were being carried by the *Radfahrer* (cyclists) of the infantry; by riflemen, sharpshooters and pioneers; by independent machine-gun units; by telegraph and field telephone units; by airship and motor transport detachments; by most cavalrymen; by foot artillerymen; and by parts of the Train.

Navy units also carried the *Karabiner* 98AZ, though distribution was limited; the first issues were made in 1914 to the *Maschinengewehr-Bedienungsmannschaften der* III. *Seebataillon* (machine-gun instructors of the third battalion of marines).

The Prussian government arsenal in Erfurt apparently made most of the *Karabiner* 98 AZ issued in Prussia, Saxony and Württemberg. However, Danzig-marked examples have also been reported. The Bavarians bought their weapons from Amberg, although production was very slow as most of the production facilities had been committed to the *Gewehr* 98; only about 30,000 Kar. 98 AZ had been delivered to the Bavarian Army by 1910.

A typical chamber mark reads 'crown/ ERFURT/1913', while 'KAR.98' will be found on the left side of the receiver immediately in front of the thumb-clearance cutaway. Serial numbers, part-numbers and inspectors' marks are similar to those of the *Gewehr* 98.

The standard bayonet issued to cavalrymen in 1914 was the S84/98; however, the foot artillery used the S98/05, with a 38cm swell-point blade, and the independent machine-gun units had the short-bladed KS98. Other accessories included a sling, a muzzle protector and a screwdriver.

Although there was still too much muzzle blast (common to all short firearms firing full-power ammunition) for the troops' liking, this was offset by the handiness of the *Karabiner* 98 AZ compared with the longer and clumsier *Gewehr* 98. Complaints were also made about the new tangent-leaf sight, which was regarded as awkward to use at maximum elevation.

Second-Rank Patterns

Small-Calibre Guns

7.9mm Gewehr Modell *1888/05* *(Gew. 88/05)*

The introduction of the *Gewehr* 98, spread over a number of years, displaced substantial quantities of *Gewehre* 88 (the *Reichsgewehr*) and the Germans were understandably reluctant to discard hundreds of thousands of serviceable weapons.

The worst feature of the *Reichsgewehr* was the clip-loaded magazine. Although this had been a great step forward in the late 1880s, the advent of the charger-loading system – which allowed guns to be replenished with single rounds – was a considerable improvement. The obvious solution was to develop an appropriate charger-loading conversion system for the Gew. 88; this resulted in production of the Gew. 88/05.

Essentially the same as the Gew. 88/S (qv), the converted weapons may be found with '•', 'Z' and 'S' chamber-top markings, which signify modified barrel contours and deep-groove rifling suited to the S-Munition.

7.9mm *Gewehr Modell* 1888/05 (Gew. 88/05)	
Synonyms:	Commission Rifle M88/05, Mauser-Mannlicher M88/05, *Reichsgewehr* M88/05
Adoption date:	3 January 1907
Length:	1,245mm (49.02in) *with bayonet* 1,747mm (68.78in) with S71
Weight:	*without sling* 3.9kg (8.6lb) *with bayonet* 4.575kg (10.09lb) with S71
Barrel length:	740mm (29.13in)
Chambering:	7.9 × 57mm, rimless
Rifling type:	four-groove, concentric
Depth of grooves:	0.15mm (0.006in)
Width of grooves:	4.4mm (0.173in)
Pitch of rifling:	one turn in 240mm (9.45in), RH
Magazine type:	single-row protruding box
Magazine capacity:	5 rounds
Loading system:	single rounds or chargers
Cut-off system:	none
Front sight:	open barleycorn
Backsight:	leaf-and-slider type
back-sight setting:	*minimum* 400m *maximum* 2,000m
Muzzle velocity:	875m/sec (2,870ft/sec)
Bullet weight:	9.85g (152gn), 'S' type

The Gewehr *88/05 showing the charger-guard block above the receiver bridge.* Ian Hogg

7.9mm *Gewehr Modell 1888/05* (Gew. 88/05)

Internal Arrangements

Though essentially similar to the Gew. 88/S, the Gew. 88/05 loads from a standard sheet-metal charger discarded automatically as the bolt closes. Among the most obvious features of the conversion is two blocks forming the charger guides, which are screwed to the top of the receiver-bridge. The left side of the receiver wall is ground-out to enable the thumb to press the cartridges fully down into the magazine well, and a semi-cylindrical channel milled vertically across the breech face allows the pointed nose of the S-*Patronen* to pass downwards into the magazine. The groove is necessary as the extra width of the charger body holds the cartridges farther forward in the magazine feed aperture than the original clip did.

The magazine was narrowed by inserting a pressed-steel strip and shortened internally by the addition of a small steel block to prevent loose cartridges rattling around the magazine well and failing to feed. A spring-loaded cartridge retainer is fixed horizontally in the left wall of the magazine well and a pressed-steel cover blocks the opening in the bottom of the magazine.

External Appearance

The Gew. 88/05 was essentially similar to the Gew. 88/S but had charger guides attached to the receiver-bridge and the modified Gew. 88/S-type backsight with a single leaf graduated to 2,000m. The small (rear) sight leaf was customarily removed and the clip-ejection slot was sealed with a tin plate cover.

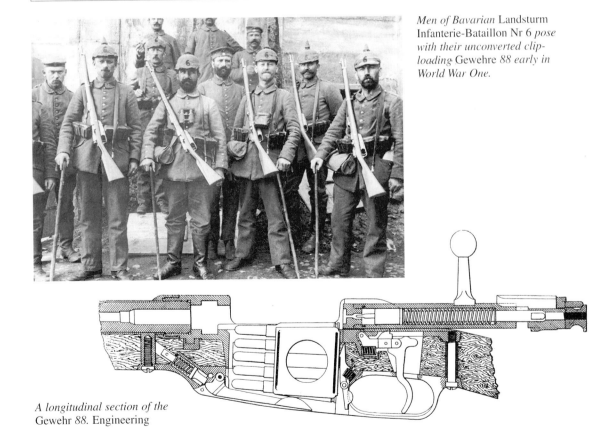

Men of Bavarian Landsturm Infanterie-Bataillon Nr 6 *pose with their unconverted clip-loading* Gewehre *88 early in World War One.*

A longitudinal section of the Gewehr *88.* Engineering

Conversion of the Gew.88/05 was apparently undertaken exclusively in the Spandau rifle factory, where about 370,000 Gew. 88 were altered in 1906–08. Stored for the active Reserve, about 70,000 were still nominally on the Army inventories in October 1918.

7.9mm Gewehr Modell *1888/S (Gew. 88/S)*

The *Gewehr* 88 *abgeändert für* S-*Patrone* was a modification of the original *Gewehr* 88 to handle the high-velocity cartridge that had been accepted on 3 April 1903. The bullet of the new cartridge had a diameter of 8.22mm, appreciably greater than the 8.1mm of the original 1888 pattern, and the rise in chamber pressures meant that only guns with newly made 'Z'-pattern barrels were altered. The chamber was adjusted to accommodate the new ammunition and the top of the receiver was marked with a large 'S'. A crowned 7mm-high 'S' was struck into the right side of the butt.

Adopted in Prussia, Saxony and Württemberg in November 1888 and in Bavaria on 19 February 1891, the Gew. 88 had been designed and introduced in less than a year to counter the French Mle 86 (Lebel) – the first military rifle to chamber a cartridge loaded with smokeless propellant. The Lebel was not a particularly noteworthy design, but the power and flat trajectory of its cartridge gave French troops a significant advantage over the Germans.

Machinery purchased from Ludwig Loewe & Cie was delivered in the autumn of 1888 to Spandau, then to Danzig and lastly to Erfurt. Deliveries of tools to Amberg commenced in July 1889, and by October the arsenals were working around the clock to make 1,000 guns daily.

The most important private contractor was to have been Waffenfabrik Mauser AG, but Mauser had obtained a huge order from Turkey and could not spare production facilities. Consequently, Ludwig Loewe & Cie entered large-scale rifle manufacture on the basis of making backsight leaves, minor components and Russian-model Smith & Wesson copies in the 1870s. However, all was not as it seemed; Loewe owned a majority of Mauser shares, and so kept business within the cartel.

7.9mm *Gewehr Modell* 1888/S (Gew. 88/S)

Synonyms:	*Reichsgewehr* M88/S, German Commission Rifle M88/S, Mauser-Mannlicher M88/S
Adoption date:	1903(?)
Length:	1,245mm (49.02in) *with bayonet* 1,545mm (60.83in) with S71/84
Weight:	*without sling* 3.825kg (8.43lb) *with bayonet* 4.185kg (9.23lb) with S71/84
Barrel length:	740mm (29.13in)
Chambering:	7.9 × 57mm, rimless
Rifling type:	four-groove, concentric
Depth of grooves:	0.15mm (0.006in)
Width of grooves:	4.4mm (0.173in)
Pitch of rifling:	one turn in 240mm (9.45in), RH
Magazine type:	single-row projecting box
Magazine capacity:	5 rounds
Loading system:	clip
Cut-off system:	none
Front sight:	open barleycorn
Backsight:	leaf-and-slider type
Backsight setting:	*minimum* 400m *maximum* 2,000m
Muzzle velocity:	875m/sec (2,870ft/sec)
Bullet weight:	9.85g (152gn), 'S' type

The original order for 300,000 rifles, dating from January 1889, was subsequently amended by the addition of 125,000. This undertaking seems to have weakened Loewe's resolve to supply machinery, as the Amberg commandant, *Oberstleutnant* Freiherr von Brandt, subsequently complained that deliveries were proceeding very slowly.

Loewe-made commission rifles cost the Prussian government substantially more than government-made items, leading to unfounded allegations of profiteering. More problems arose in

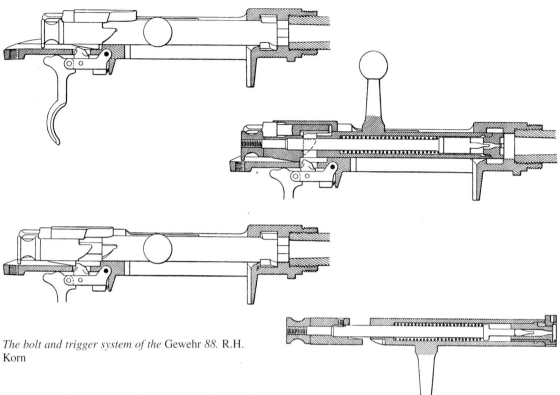

The bolt and trigger system of the Gewehr 88. R.H. Korn

October 1889, when the Prussians tried to order 300,000 rifles from Österreichische Waffenfabrik-Gesellschaft and all work stopped while legal wrangling over infringement of patents granted to Mannlicher and Mieg was resolved (*see* Introduction).

Issues of the *Gewehr* 88 began in the autumn of 1889 to XV. and XVI. *Armeekorps*, then stationed in Elsass-Lothringen. The first large-scale issues in Bavaria seem to have commenced on 22 October 1889, although small numbers had previously been used for field trials. The three Prussian arsenals had delivered about 275,000 Gew. 88 by the beginning of 1890, allowing all the line infantry regiments of Prussia, Saxony and Württemberg to be re-equipped with the new rifles by 1 August.

By the end of 1890, weapons were beginning to come from Amberg, Loewe and Österreichische Waffenfabrik-Gesellschaft, and possibly also from private companies in Suhl. However, although Schilling and Haenel are known to have made

Karabiner 88, *Gewehre* 91 and sporting rifles on the *Reichsgewehr* action, no *Gewehr* 88 has yet been reported from these sources.

Production finally ceased in 1897. At least 750,000 Gew. 88 had been made in the Prussian government arsenals in Danzig, Erfurt and Spandau; about 100,000 had come from the Bavarian government arsenal in Amberg; 425,000 were supplied by Ludwig Loewe & Cie of Berlin-Charlottenburg; 300,000 by Österreichische Waffenfabriks-Gesellschaft in Steyr; and perhaps 100,000 carbines and short rifles had emanated from contractors in Suhl.

From 1900 onward, Gew. 88, displaced by new Mausers, were relegated to the Reserve, *Landwehr* and *Landsturm*. In 1906, however, the first of about

7.9mm *Gewehr Modell* 1888/S (Gew. 88/S)

Internal Arrangements

The Gew. 88 is a simplification of the M1871/84 (Mauser) pattern, retaining the trigger, split-bridge receiver and many other features of the earlier weapon. The receiver is a one-piece forging with the bolt-handle locking down in front of the bridge. A small bolt-stop – based on earlier Mauser designs – lies on the rear left side of the receiver, where it can pivot outwards to release the bolt.

The bolt is a slender forged tube with an integral handle terminating in a ball. The spring-loaded striker slides inside the hollowed bolt body, shoulders towards the rear providing a bearing for spring compression; the cocking-piece and the safety mechanism are attached to the striker extension.

The bolt body carries the two locking lugs, one solid and the other slotted longitudinally to allow a finger on the bolt-stop to ride into the bolt and operate the ejector. The solid (bottom) lug is bevelled to match the contours of the face of the receiver ring and gives adequate primary extraction when the bolt-handle is raised. The bolt lugs lock vertically in the solid receiver ring, immediately behind the chamber.

The detachable bolt head is one of the poorer features of the Gew. 88 and can easily be lost. It fits closely inside the front of the bolt body, relying on a lug on the sub-diameter rearward extension to match a recess cut in the bolt in such a way that it can only be removed in one particular position. The extractor, a weak spring-steel claw, slides in a groove cut in the right side of the bolt body. A small supplementary screw was sometimes added to retain the extractor, but the claw was widened when the bolt face was redesigned in 1891.

The bolt head face was originally circumferentially recessed for the cartridge head, but the wall was subsequently partly cut away to prevent double loading. The bolt head is prevented from rotating by flats on the striker head.

A standard Mauser-type 'wing' safety mechanism lies on top of the cocking-piece. Raised vertically, it cams the cocking-piece out of engagement with the sear and locks the bolt shut. The sear and trigger units are simple, strong and arranged to give a two-stage pull.

The trigger-guard and magazine case are formed as a single unit, the magazine being loaded with a five-round Mannlicher-inspired clip adapted by Gewehr-Prüfungs-Kommission technicians to feed either way up. The clip falls downwards out of the magazine well when the last round has been stripped into the chamber.

Cartridges are delivered to the feed way by a follower arm, propelled by a spring-loaded pin in the front of the magazine body. A spring-loaded latch, protruding into the front of the trigger-guard bow, retains the clip when the bolt is open. Pressing the catch-head once the action has been opened throws the clip and its contents up and out of the gun.

External Appearance

The Gew. 88 has a horizontal bolt-handle, locking down ahead of the receiver-bridge, and a straight one-piece walnut stock with a reinforced butt-toe. The stock sweeps gracefully up to the magazine (formed integrally with the trigger-guard bow) that protrudes below the stock underneath the bolt mechanism. A large-diameter sheet-steel jacket protects the barrel, and a single spring-retained barrel band carries the front sling swivel. A small bayonet lug lies on the right side of the simple nose-cap and the head of a half-length cleaning rod protrudes beneath the muzzle.

The rear sling swivel lies beneath the butt, with an alternative anchor position in the front face of the magazine to shorten the sling for parade use. The small leaf of the original 1888-type sight was abandoned, a new 2,000m sight leaf was substituted for the old 2,050m pattern, and the standing block battle sight sufficed for 400m rather than 250m. Some writers have suggested that new sight leaves were supplied for the Gew. 88/S, but others insist that the markings on the old sight leaves were simply replaced. Both claims may be partly true.

Operation

Starting with the action locked on a fired case, the bolt-handle is raised vertically. The rotary motion forces the cocking-piece backwards, compressing the striker spring, and the camming action of the locking lugs in their seats loosens the spent cartridge case. Mechanical advantage reduces the effort considerably.

Drawing the bolt-handle backwards withdraws the spent case, held by the extractor claw, until it is thrown from the feed way by the ejector. As the bolt-handle is pushed forward again, a new cartridge is stripped from the clip and thrust into the chamber. The final seating is achieved as the bolt-handle is turned down, gently camming the cartridge forward. The gun is then loaded, cocked and ready to fire. If no more cartridges remain in the magazine, the empty clip falls down and out through the floor plate aperture.

370,000 rifles were converted to charger-loading *Gewehre* 88/05 (qv). Many *Reichsgewehre* were sold to China in 1907 and, even as late as 1911, A.L. Frank of Hamburg – 'Alfa' – was still trying to sell 40,000 of them.

Unaltered Gew. 88 were also sold in large numbers in the Balkans and the Far East. Österreichische Waffenfabrik-Gesellschaft exported guns to Peru and Brazil in the early 1890s, and some were even used during World War One by the Austro-Hungarian Army.

The best feature of the *Gewehr* 88 was its superiority over the Lebel and the earliest Austro-Hungarian Mannlichers. However, parts of the basic design were poor and its introduction excited great controversy (*see* Part One for a fuller account).

The *Patrone* 88, loaded with 2.75g of *Gewehr-Blattchen-Pulver* 88, had a muzzle velocity of about 620m/sec and a chamber pressure of 3,305kg/cm². The round-nose bullet weighed 14.5g. Performance included an ability to penetrate a 7mm-thick iron plate at 300m, or, alternatively, 80cm of pine at 100m, 45cm at 400m and 25cm at 800m. Maximum range was found to be about 3,800m at an elevation of 32 degrees, and the 'radius of dispersion' was reckoned to be 0°38′25″ at 500m.

Explosions in the magazine were traced to double loading, when the extractor failed to slip into the extraction groove as the cartridge was pushed into the chamber. If the firer attempted to reload,

The original Patrone *88, bulleted, blank and drill cartridges.*

(Above, left to right) *The* Patrone *88, a longitudinal section of the cartridge case, and the original bullet drawn to a larger scale.*

(Right, left to right) *The standard blank, its wooden bullet* (Holzgeschoss), *and two types of one-piece drill cartridges.*

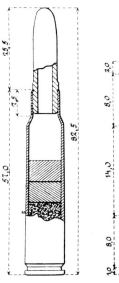
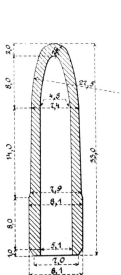
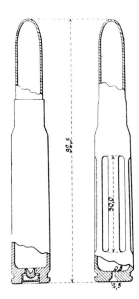

the nose of the second round was struck into the primer of the first cartridge that had been left in the chamber. If the blow was sufficiently hard, the chambered cartridge exploded. The chances of serious injury were high, as the open bolt could be slammed back with tremendous force.

The problem was largely cured by cutting away part of the recessed bolt head face so that cartridges rose out of the magazine directly under the extractor claw, instead of expecting the claw to slip over the case rim as the bolt closed.

Explosions in the chamber, mostly due to badly loaded ammunition, were reduced by improving quality control. In addition, from 9 January 1891, the barrel was strengthened by the substitution of a straight-sided cone for the original concave taper directly in front of the chamber. This greatly reduced rupturing incidents, but did not cure the problems entirely. The Bavarian Army alone returned more than 1,000 badly damaged barrels to Spandau in 1900–01.

Adding two gas-deflecting lugs on the striker head from 1894 eased excessive gas leakage arising from pierced primers and flaws in cartridge-case heads. The large lug on the left side projected into the locking-lug guide-way, while the smaller (under) lug blocked the cocking-cam guide-way. All service rifles were subsequently altered, and the original circular striker heads are now very rarely seen.

The Germans realized as early as 1890 that the rifling design pirated from the Lebel had serious shortcomings, as the lands eroded with alarming rapidity. A check of all *Gewehre* 88, *Karabiner* 88 and *Gewehre* 91 in store in the summer of 1893 revealed that half needed new barrels, even though few of them were more than three years old.

Tests made by the *Gewehr-Prüfungs-Kommission* in 1894 provided the answer. The 8.1mm diameter of the standard bullet had to be reduced to a bore diameter of 7.9mm, but a groove depth of 0.1 mm squeezed the bullet too greatly. Rapid bore wear was encouraged by excessive friction between the bullet jacket and the bore walls, and chamber pressures increased to levels which were often dangerous.

Bullet jackets tended to crack under pressure, and excessive metal fouling was deposited in the barrel.

The *Gewehr-Prüfungs-Kommission* showed that 0.15mm grooves gave a much better combination of bore life, accuracy and chamber pressure, and appropriately modified rifling was adopted on 7 July 1896. Guns with new-pattern rifling were marked with a 3mm letter 'Z' on top of the chamber, and sometimes also with a 7mm 'Z' on the right-hand side of the butt.

Service experience soon revealed the shortcomings of the barrel jacket. Despite protecting the barrel surface and insulating the firer from excessive barrel heating, it was easily dented, rusted at the joints, and could be sprung out of alignment with a consequent and undesirable effect on accuracy.

The split-bridge receiver was also disliked, although the sloppiness often evident as the bolt was retracted was not particularly evident in the well-made *Gewehr* 88. Minor flaws originally included a slow striker fall or 'lock-time', owing to the excessive weight of the striker head, cocking-piece and safety unit; a detachable bolt head which could easily be lost; and, prior to 1891 at least, a weak extractor claw.

The clip-loading system prevented the magazine from being replenished with single rounds, which, though it seemed a great advance in the 1880s, was soon seen to be undesirable on the battlefield. Many *Gewehre* 88 were converted to charger loading in 1906–07 (Gew. 88/05) and 1914–15 (Gew. 88/14).

Manufacturers' marks are usually found on top of the receiver above the chamber, accompanied by the production date. Most of the factories are easy to identify, though Österreichische Waffenfabriks-Gesellschaft of Steyr customarily used nothing but an 'OEWG' mark. The designation 'GEW. 88' is stamped into the rear left side of the receiver, while the serial number – sometimes with a suffix letter – appears on the left side of the receiver, on the barrel clamping ring alongside the breech, and on the base of the bolt-handle.

Stampings above the chamber may include a 2mm-diameter '•', applied after 1891 to indicate modified barrel contours; 'Z' (3mm high), for deepened post-1896 rifling grooves; and 'S' (3mm high), showing that the chamber had been altered for S-Munition in 1903–05. A 7mm 'Z' or crowned 'S' may be present on the right side of the butt.

A Fraktur 'n m' mark beneath the 'GEW. 88' designation on the left rear of the receiver of many surviving guns still warrants a suitable explanation. It has been suggested that *nitro-munition* identified guns which survived the introduction of cartridges loaded with *Pulver* 436 (1898). But these were dimensionally identical to their predecessors and it is hard to see how the reduced chamber pressure could have justified distinguishing the guns from those that had been firing ammunition loaded with *Gewehr-Blattchen-Pulver*.

An alternative explanation, that 'n m' represents *nachgeschnittener Modell* (indicating that the rifling grooves of existing barrels had been deepened), is worthy of consideration.

The bayonet originally issued with the *Gewehr* 88 to the infantrymen of Prussia, Saxony and Württemberg was the brass-hilted *Seitengewehr* 71, which had a 47cm sword blade. The *Jäger* (riflemen) retained the *Hirschfänger* 71, which had a 50cm blade and a leather-gripped steel hilt, but Bavarian infantry units preferred the 25cm-bladed S71/84.

Other accessories included two patterns of cleaning rod – *Wischstöcke* 88 and 93 – and a muzzle protector. Screwdrivers were issued on the scale of one to every ten guns and lock spanners on the scale of one to every three. There was also a sling.

7.9mm Karabiner Modell *1888/S (Kar. 88/S)*

These were the survivors of the short-barrel guns made prior to 1900. A few had been converted for the S-Munition in 1903–05 and bore an additional chamber-top 'S'. However, because of its short barrel, the Kar. 88/S – and its near relation, the *Gewehr* 91/S (qv) – suffered from unpleasant muzzle blast and flash even with the original *Patrone* 88. Using more powerful S-Munition emphasized the problems. The *Gewehr-Prüfungs-Kommission* had begun development of a shortened version of

---

7.9mm *Karabiner Modell* 1888/S (Kar. 88/S)	
Synonyms:	*Reichskarabiner* M88/S, Commission Carbine M88/S, Mauser-Mannlicher carbine M88/S
Adoption date:	19 January 1890
Length:	955mm (37.6in)
Weight:	3.125kg (6.89lb) without sling
Barrel length:	435mm (17.13in)
Chambering:	7.9 × 57mm, rimless
Rifling type:	four-groove, concentric
Depth of grooves:	0.15mm (0.006in)
Width of grooves:	4.4mm (0.173in)
Pitch of rifling:	one turn in 240mm (9.45in), RH
Magazine type:	single-row projecting box
Magazine capacity:	5 rounds
Loading system:	clip
Cut-off system:	none
Front sight:	open barleycorn
Backsight:	leaf-and-slider type
Backsight setting:	*minimum* 250m *maximum* 1,200m
Muzzle velocity:	775m/sec (2,545ft/sec)
Bullet weight:	9.85g (152gn), 'S' type

---

the new infantry rifle early in 1889, completing work by the autumn. Prussia, Saxony and Württemberg adopted the *Karabiner* 88 at the beginning of 1890. Ironically, issue of single-shot M1871 carbines to the cuirassiers (the last cavalrymen to receive them) was completed in the same year.

The carbines are mechanically identical to the standard rifles, differing only in size and fittings. They resemble the standard *Gewehr* 88 but are considerably shorter, have turned-down bolt-handles, are stocked to the muzzle and lack bayonet lugs.

The plain nose-cap has projecting 'ears' guarding the front sight. The spring-retained barrel band has a fixed sling bar on the left side; another sling-anchor point is cut through the butt. A small leaf-pattern backsight lies on the barrel jacket immediately ahead of the barrel-clamping ring.

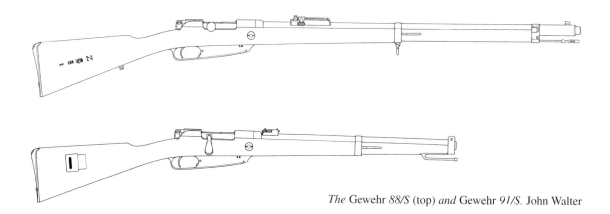

The Gewehr *88/S* (top) *and* Gewehr *91/S.* John Walter

The 1888-system carbine suffered from many of the faults discovered in the *Gewehr* 88 (qv), which included rupturing of the barrels, excessive gas leakage from damaged primers or case-heads, and excessive rifling wear. Modifying the barrel contours from January 1891 prevented ruptured barrels, and excessive gas leakage from ruptured primers of defective cartridges was minimized after 1894 by fitting new striker heads with gas deflection flanges. Excessive bore-wear was answered in 1896 by deepening the grooves in barrels.

Contracts for 200,000 guns for Prussia, Saxony and Württemberg, placed with the firearms manufacturers in Suhl – principally Haenel and Schilling – are believed to have been completed in 1892. The first contracts must have been placed in 1889, prior to the official adoption of the carbine in January 1890; sufficient guns had been delivered by March 1890 to permit issue to each Prussian cavalry squadron to begin. Apparently the Kar. 88 was adopted in Bavaria in 1891.

Hans-Dieter Götz, in *Die deutschen Militärgewehre und Maschinenpistolen 1871–1945*, records that 24,160 Kar. 88 were on hand in Bavaria in May 1893; Prussia, Saxony and Württemberg then had about 75,000 between them. Production in the Erfurt arsenal (begun in small numbers in 1891) ceased in 1896 after as many as 300,000 had been made.

Karabiner 88 served dragoons, hussars, cuirassiers and lancers, as well as the Train and some other specialized units, but were superseded by the *Karabiner* 98 AZ with effect from January 1908. The last survivors were not withdrawn from regular units until 1910, going into store or to arms dealers; the 1911 catalogue of A.L. Frank Exportgesellschaft recorded 8,200 awaiting sale.

Surviving Kar. 88 will be found with markings applied on behalf of munitions columns and other minor formations that were not likely to use the guns in anger, but most of these date from World War One.

Manufacturers' marks were struck above the chamber, typically reading 'C.G. HAENEL/ SUHL/1891', with the designation mark 'KAR.88' in Fraktur on the left side of the receiver. Serial numbers appear on the left side of the receiver, on the barrel-clamping ring alongside the breech, and on the base of the bolt-handle. Stampings above the chamber may include '•', 'S' and 'Z'.

The carbines did not initially accept a bayonet; however, a few guns are now encountered with auxiliary bayonet lugs. These *Zollkarabiner* 88 were apparently intended for the German customs service. Kar. 88 accessories included a sling, a screwdriver, a lock spanner and a leather muzzle protector. There were also several different saddle scabbards (for example, *Futteräle* 88, 95, 96).

7.9mm Gewehr Modell *1891/S (Gew. 91/S)*

This carbine-length rifle was adopted for the foot artillerymen of Prussia, Saxony and Württemberg

7.9mm *Gewehr Modell* 1891/S (Gew. 91/S)

Data: generally similar to *Karabiner* 88/S, but a little heavier.

on 25 March 1891, but was also issued to the German Army's foot artillerymen and specialized detachments such as the *Luftschiffer* units. The Gew. 91 was little more than a Kar. 88 with an additional stacking rod made integrally with a steel plate let into the underside of the fore-end immediately behind the nose-cap. It shared the same major shortcomings as the Kar. 88: excessive muzzle blast and flash.

The bolt-handle is turned downward, a plain nose-cap has 'ears' protecting the sight, the barrel band has a sling bar on the left side, and a sling-anchor is cut through the butt. A small leaf-type backsight is situated on the barrel ahead of the chamber.

Gewehre 91 bear their maker's marks over the chamber, in typical German fashion; production was never large, though more than 18,000 had been delivered to Bavaria alone by May 1893. Most surviving examples bear inscriptions such as 'C.G. HAENEL/SUHL/1892', or 'V.C. SCHILLING/SUHL/1894'. The designation mark 'GEW.91' in Fraktur may be found on the left side of the receiver. Proof marks will be found on the left side of the barrel and the receiver. Serial numbers appear in full on the left side of the barrel and the receiver, and on the base of the bolt-handle.

Some guns were apparently fitted with bayonet-attachment bars during World War One, when it is assumed that the piling hook was removed. The designation stamp on the receiver would then have been the only way to distinguish these *Gewehre* 91 from similarly modified *Karabiner* 88.

Each Gew. 91 was accompanied by a sling, a leather muzzle protector, a screwdriver (on a scale of one to every ten guns) and a lock spanner (one to every three guns).

Large-Calibre Types

11mm Infanterie-Gewehr Modell *1871/84 (Gew. 71/84)*

The *Gewehr* 71/84 was adopted in Prussia, Saxony and Württemberg at the beginning of 1884, but the obsolescence of the existing production facilities – which required extensive renewal – delayed production work until the summer of 1886. The Bavarian *Kriegsministerium* was given an example of the perfected magazine rifle on 8 May 1884

11mm *Infanterie-Gewehr Modell* 1871/84 (Gew. 71/84)

Synonym:	Mauser M1871/84
Adoption date:	31 January 1884
Length:	1,295mm (50.98in) *with bayonet* 1,765mm (69.5in) with S71
Weight:	*without sling* 4.6kg (10.14lb) *with bayonet* 5.025kg (11.08lb) with S71
Barrel length:	800mm (31.5in)
Chambering:	11×60mm, rimmed
Rifling type:	four-groove, concentric
Depth of grooves:	0.3mm (0.012in)
Width of grooves:	4.5mm (0.177in)
Pitch of rifling:	one turn in 550mm (21.65in), RH
Magazine type:	under-barrel tube
Magazine capacity:	8 rounds
Loading system:	single rounds
Cut-off system:	none
Front sight:	open barleycorn
Backsight:	a leaf and a leaf-and-slider attached to a standing block
Backsight setting:	*minimum* 200m *maximum* 1,600m
Muzzle velocity:	435m/sec (1,425ft/sec)
Bullet weight:	25g (386gn)

Note: the depth of the rifling grooves was reduced to 0.15mm (0.006in) on rifles made after 14 November 1885 in Prussia, Saxony and Württemberg, and after 1886 in Bavaria.

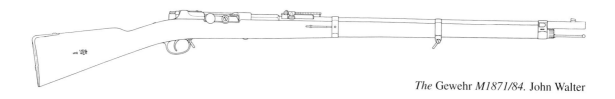

The Gewehr *M1871/84*. John Walter

11mm *Infanterie-Gewehr Modell* 1871/84 (Gew. 71/84)

Internal Arrangements

The *Gewehr* 71/84 is similar mechanically to the M1871 (qv), apart from the changes necessary to accommodate the cartridge elevator and the tubular magazine. Although the upper part of the receiver is very similar to its predecessor, the forging has been deepened to provide a hollow box for the cartridge elevator mechanism. Major adjustments have also been made to the 'wing'-type safety lever, the trigger assembly has been completely redesigned, and the extractor claw has been moved to the top right side of the bolt head.

The 'U'-trough elevator is operated by a cam lying in a recess milled in the left wall of the box, which acts in conjunction with the spring-loaded magazine cut-off lever. The knurled head of the cut-off protrudes from the left rear of the receiver alongside the bridge. Pulled back, the cut-off raises the elevator-operating cam into the ejector channel in the left side of the bolt-way; pushed forward, the cut-off lowers the elevator cam to isolate the elevator from the operating stroke of the bolt.

The elevator is operated by an ejector rib, retained by a spring collar, which runs almost the entire length of the left side of the bolt body. A groove milled along the lower edge of the rib raises the cartridge elevator as the bolt reaches the end of its backward travel, and depresses it again as the bolt nears the end of the closing stroke. The front end of the ejector protrudes through a slot in the bolt head to kick the cartridges out of the bolt way as the ejector rib is stopped by the elevator-operating cam – or, alternatively, by a stud on the cut-off spring if the cut-off has been applied.

A spring-loaded stop in the left side of the elevator box holds the cartridges in the magazine when the elevator is in its raised position. The stop is released as the elevator drops, allowing a new round to be pushed backwards out of the magazine tube.

External Appearance

The M1871/84 has a bolt-handle that locks horizontally down in front of the receiver-bridge, and a one-piece stock with a straight wrist. The trigger-guard, an integral part of the floor plate, has a hole through the front web to accept a sling swivel. The back band is held by a leaf spring, the middle band carries a sling swivel, and the plain nose-cap has a bayonet lug on its right side. The tubular magazine protrudes below the muzzle, where it terminates in a robust cap with a straight ball-tipped stacking rod. The backsight is mounted on the barrel in front of the short octagonal reinforce. The original infantry type was eventually replaced by the so-called *Jäger-Visier*, which is graduated differently.

Operation

The M1871/84 rifle is used in much the same way as the single-shot M1871, except that the tube magazine can be loaded through the top of the open action and a cut-off lever lies on the rear left side of the receiver.

---

and King Ludwig II signed adoption papers soon afterward.

New production machinery purchased from Ludwig Loewe & Cie of Berlin, Germany's premier machine-tool manufacturer, was installed in the government small arms factories during 1885. A cold-drawing process – perfected in the Danzig factory for tubular lance-bodies – was adapted to make the magazine tubes. Engineering problems were gradually overcome, and the first new rifles were delivered from Spandau in March 1885. By the beginning of 1886, Erfurt, Danzig and Amberg had all made their initial contributions. Württemberg ordered rifles directly from Waffenfabrik Mauser AG, 19,000 being delivered in the summer of 1886.

Almost 100,000 *Gewehre* 71/84 were delivered to Saxony, Württemberg and the *Kaiserliche Marine* in September 1887 alone. By the end of the year, the inventory in Bavaria stood at 96,368 *Gewehre* 71/84 and it is suspected that total production from the beginning of 1885 until early 1889 (when assembly ceased) approached one million.

The first issues of magazine rifles were made in July 1886 to the Prussian XV and XVI army corps, guarding the borders in Elsass-Lothringen. These men had been prepared for the new rifles by the distribution of about 1,000 *Demonstrations-schlosse* or sectioned actions.

As so often happens with weapons, however, faults became apparent almost as soon as the Gew. 71/84 entered service. Alterations had to be made to the machining of many components, and trials undertaken by the *Jäger-Bataillone* in 1885–86 showed that the rifle shot consistently to the right. The front sight was subsequently offset to the right to correct the point of impact, and armourers were authorized to increase lateral deviation by as much as 1.2mm to rectify rogue rifles.

Shooting was sensitive to barrel heating and changes in atmospheric temperature. This was blamed on variations in the point of balance as the cartridges were expended, and by the warping of many hastily seasoned stocks.

The Gew. 71/84 was difficult to dismantle and could be jammed by dirt and dust; and the cut-off lever often broke away, leaving the firer with a useless magazine if the cut-off had been set for single-shot fire. Yet the rifle remained popular with the rank-and-file in spite of its weight.

The advent of the French Lebel rifle made the Mauser obsolescent overnight, dealing a great blow to German prestige as series production of the Gew. 71/84 had barely begun. Although the Lebel was also built around a tube magazine, the 8mm French cartridge was loaded with smokeless propellant and developed a muzzle velocity about 40 per cent greater than the ponderous 11mm *Reichspatrone*.

The Germans developed the *Gewehr* 88 (qv) to counter the Lebel, but, in the panic of the moment, came close to accepting an 8mm version of the *Gewehr* 71/84 with an additional lug. Known as the *Mauser Kammer mit doppeltem Widerstand* (Mauser bolt with double locking lugs), this system had been patented in Austria-Hungary in September 1887. A new lug on the rear of the bolt body revolved into a recess milled in the left wall of the receiver behind the cartridge-elevator box, giving additional strength in an action normally locked only by the abutment of the bolt guide rib on the receiver-bridge.

The two-lug bolt was not adopted in Germany, owing to progress with smokeless propellant and small-calibre cartridges, but was incorporated in the 1887-model Turkish rifle (qv).

Gewehre 71/84 were gradually withdrawn as the more efficient 1888-pattern rifles entered service. Many of the black-powder weapons were relegated to the Reserve and the *Landwehr*, while others were sold as unwanted surplus for a few Marks apiece; in 1911, A.L. Frank of Hamburg still had 12,000 in stock. Those that could be retrieved in 1914 were reissued to lines of communication, *Landwehr*, *Landsturm* and recruiting-depot personnel.

A few Gew. 71/84 are said to have been chambered for 8mm cartridges, by boring-out the barrels and inserting a new rifled liner. This is believed to have been done during the early 1890s to offset temporary shortages of Gew. 88, but no such guns could be traced for examination.

Although the front-line service life of the Gew. 71/84 was short, guns displaced from the Army served the Navy for many years. Six thousand Gew. 71/84 were being used by I. *Matrosen-Division* in 1905, when nearly 10,000 were shared amongst the four *Matrosen-Artillerie-Abteilungen*. Five hundred Gew. 71/84 were issued to the two

Torpedo-Abteilungen, and others went to *Augmentationsschiffe* (auxiliary warships).

Substantial numbers of these obsolescent rifles were still in the hands of naval personnel when World War One began; indeed, issues to land-based units had been increased after 1910 to free 1898-type Mausers to arm the newest warships.

A manufacturer's mark appears on top of the barrel octagon ahead of the receiver ring. The Prussian arsenals in Danzig, Erfurt and Spandau, the Bavarian arsenal in Amberg and Waffenfabrik Mauser AG of Oberndorf are all known to have been involved. The designation 'I.G.MOD.71/84' in Fraktur lies on the left side of the receiver, with the date of manufacture on the right side of the bridge. Two dates usually signify that manufacture and acceptance occurred in differing years.

The Gew. 71/84 was officially issued with the S71/84, a short knife bayonet with a 25cm blade. However, the Prussian guard infantry retained the long S71 and *Jäger* units kept the *Hirschfänger* 71. There was also a sling, a muzzle protector, a backsight cover and suitable cleaning equipment.

11mm Modell *1871/84 Short Rifle*

It was normal German practice to develop a cavalry carbine and a *Jägerbüchse*, for the riflemen, in addition to infantry rifles. However, the restricted career of the Gew. 71/84 rifle prevented short weapons being introduced in quantity, though carbines were made experimentally and a few thousand short rifles were made for trials with the *Jäger-Bataillone* in 1885–86.

Some guns survived long enough to be issued to regimental depots and similar non-combatant units during World War One. A typical short rifle, made in Spandau in 1886, is about 60mm shorter than normal. The magazine tube protrudes farther below the muzzle than the Gew. 71/84 type, the magazine cap has a short ball-tipped piling rod, and a recoil bolt runs through the stock below the chamber. The bayonet is assumed to have been the *Hirschfänger* 71; although the marking on the butt suggests issue to a second-line unit, it is clearly a twentieth-century addition.

The Gewehr *71/84 remained in the German Navy in large numbers long after it had been discarded by the Army. This sailor from SMS* Goeben *was pictured shortly before World War One began.*

Single-Shot Types

11mm Infanterie-Gewehr Modell *1871 (Inf.-Gew. M71)*

The M1871 Mauser was among the earliest bolt-action weapons to be developed satisfactorily, as the gun patented in 1867–69 was substantially the same as that adopted in 1872. The major differences lay in details instead of the concept.

Several problems arose when the M1871 entered service. Poor accuracy was ultimately traced to the design of the muzzle and the bayonet attachment assembly. Very little was known in the 1870s about the effects of harmonics set up in the barrel during firing, and most barrels were held in the stock by clamping barrel bands which could exert a critical effect on accuracy.

The Gewehr *M1871.*

**11mm *Infanterie-Gewehr Modell* 1871
(Inf.-Gew. M71)**

Synonym:	Mauser rifle M1871
Adoption date:	22 March 1872

Data from a gun tested in Britain in 1876

Length:	1,341mm (52.8in)
	with bayonet
	1,842mm (72.5in) with S71
Weight:	*without sling*
	4.65kg (10.25lb)
	with bayonet
	5.33kg (11.75lb) with S71
Barrel length:	850mm (33.46in)
Chambering:	11 × 60mm, rimmed
Rifling type:	four-groove, concentric
Depth of grooves:	0.4mm (0.01in)
Width of grooves:	4.5mm (0.177in)
Pitch of rifling:	one turn in 550mm
	(21.65in), RH
Loading system:	single rounds
Front sight:	open barleycorn
Backsight:	a small leaf and a large
	leaf-and-slider attached to
	a standing block
Backsight setting:	*minimum* 300m
	maximum 1,600m (with
	sliding extension)
Muzzle velocity:	435m/sec (1,427ft/sec)
Bullet weight:	25g (386gn)

Pictured in 1915, this Landsturmmann *poses with a* Gewehr *M1871.*

Variations in humidity, barrel heat, and many other factors could alter the bedding pressure – or the contact areas between the barrel and the stock – to change bullet trajectory. Even a fractional change of direction at the muzzle could cause considerable deviation at the target.

The method of attaching the M1871 nose-cap to the barrel was poor. The nose-cap was such a close fit over the wooden fore-end that, if the stock swelled slightly on a humid day, the pressure of wood on metal could force the nose-cap fractionally out of line (in almost any direction) and thus

---

11mm *Infanterie-Gewehr Modell* 1871 (Inf.-Gew. M71)

Internal Arrangements

The M1871 receiver is bored to accept the bolt, slotted to permit the bolt-handle to pass through the bridge, and cut away on the right side to facilitate loading. A prominent abutment on top of the rear part of the receiver acts as a bolt-stop. The barrel screws into an octagonal reinforce at the front of the receiver body.

The bolt consists of the head, the body and the cocking-piece assembly. The removable bolt head is held in the body by a small lug engaging in a recess cut in the underside of the guide rib extending forward from the base of the bolt-handle. The extractor claw, attached to the left side of the bolt head, rides in a slot cut in the receiver wall to prevent the head rotating as the bolt-handle is raised. A forked rearward extension on the bolt head engages flats on the striker nose to prevent the striker and striker head rotating.

The hollow bolt body, containing the striker and its spring, carries the bolt-handle and a prominent bolt-guide rib. The rear of the body has a short cut-out to engage the stem of the safety lever – mounted in the cocking-piece body – and a cam slot to move the cocking-piece and striker assemblies backwards as the bolt-handle is raised.

The cocking-piece, the safety lever and the striker-retaining nut form a single sub-assembly. An extension on the cocking-piece runs forward between the sides of the bolt-way to prevent the cocking-piece turning as the bolt is operated; and a triangular projection on the front right side of the cocking-piece body meshes with the cam slot in the rear of the bolt cylinder to withdraw the striker into the bolt. A cylindrical safety spindle with a 'wing' head, on top of the cocking-piece, can be revolved to lock the bolt and striker.

External Appearance

The M1871 is long and clumsy, with a prominent octagonal reinforcement at the breech. The barrel is held in the one-piece stock by two sprung bands and a nose-cap. A transverse bolt running through a reinforcing block to the underside of the muzzle attaches the nose-cap. A bayonet lug lies on the right side of the nose-cap body, a short section of the muzzle crown being turned to a smaller diameter to accept the muzzle ring. One sling swivel lies under the middle band and another is attached to the brass trigger-guard. The leaf-type backsight lies immediately ahead of the junction of the octagonal reinforce and the cylindrical part of the barrel.

Operation

Starting with the gun in its fired condition, the bolt-handle is raised and the camming action of the bolt-guide rib on the angled breech face begins extraction and simultaneously withdraws the nose of the striker into the bolt. When the bolt reaches the vertical position, the rear of the guide rib can slide back through the split portion of the receiver behind the feed way. The bolt can be retracted until the prominent annular bolt stop on top of the guide rib halts against the projecting abutment on the receiver-bridge. The spent case is removed from the feed way manually – there is no mechanical ejector – and a new cartridge can be inserted into the chamber. Closing the bolt and turning the handle down gently seats the cartridge. As the cocking-piece is being held back on the sear, the gun can be fired again.

---

ruin the accuracy of fire. Firing the rifle with the bayonet in place further impaired accuracy, as did the asymmetrical locking system. The problems were never really overcome during the M1871's service life but, fortunately, were never bad enough to be considered seriously degrading to its performance.

Strengthening the mainspring and revising the design and metal thickness of the primer walls rectified erratic cartridge ignition. Alterations made in 1882 included the addition of a tiny coil-spring to the hinge of the small backsight leaf; the addition of a hardened bolt extension-piece, pinned in the rear of the guide rib to prevent excessive wear in the locking shoulder; and a new retaining pin in the appropriate retaining screw to prevent loss of the annular bolt-stop.

The first M1871 rifles, made largely by hand in the Spandau factory in 1872, bore a spread eagle, 'SPANDAU' and '1872' on the right side of the

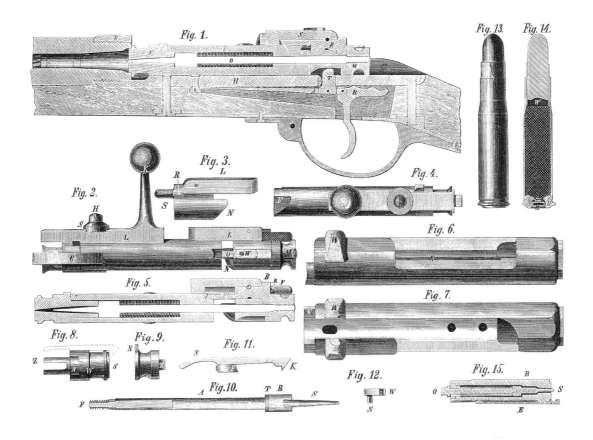

Drawing of the M1871 rifle from Schott's Grundriss der Waffenlehre *(1876).*

receiver behind the feed way. Later guns had the maker's mark on top of the barrel octagon. They were made by the Prussian government arsenals in Danzig, Erfurt and Spandau, by the Bavarian arsenal in Amberg, and by private contractors such as Greenwood & Batley Ltd, Arbeitsgemeinschaft C.G. Haenel, Gebrüder Mauser & Cie, Waffenfabrik Steyr (Österreichische Waffenfabriks-Gesellschaft), the National Arms & Ammunition Co., Ltd, Spangenberg & Sauer and V.C. Schilling & Cie.

The designation 'I.G.MOD.71' is stamped on the left side of the receiver, in Fraktur, with the date of manufacture on the right. Two dates are occasionally encountered, notably on rifles issued in Bavaria; the first is the date of manufacture, the second represents the year of issue.

A serial number will be found on the left side of the barrel octagon and on the left front side of the receiver. There may also be a calibre mark on the left side of the barrel octagon, immediately in front of the joint with the receiver (for example, '10,95' for 10.95mm).

The 1871-pattern rifle was originally issued with the *Seitengewehr* M71, which had a 47cm sword blade, a cast-brass hilt, and a wrought iron or steel cross-guard containing a muzzle ring. Other accessories included a cleaning rod, sling, muzzle protector, sight cover, screwdriver, spare striker spring and extractor tool.

11mm Jägerbüchse Modell *1871 (JB Mod. 71)*
Once the first issues of the Mauser infantry rifle had been made in 1875, the élite riflemen, the *Jäger-Bataillone*, began to clamour for something to replace the needle-fire *Jägerbüchse* M1865. Trials began as soon as the *Infanterie-Gewehr* M1871 had been approved.

The action of the *Jägerbüchse* M1871 is identical to that of the infantry rifle. The barrel is about 100mm shorter, but retains an octagonal reinforce at the breech and a turned-down muzzle crown accepting the cross-guard ring of the bayonet. The iron trigger-guard bow has a long finger-rest extending backwards and the barrel band has a sling swivel; the other swivel lies beneath the butt. The *Jäger-Visier* backsight leaf is graduated for 1,150m and 1,250m instead of the customary 1,100m and 1,200m.

By the end of 1876, the riflemen had been entirely re-equipped and the 1865-type Dreyse had been handed to the *Landwehr*. The *Jäger-Bataillone*, however, soon voiced complaints about their new Mauser, as the standard trigger was inferior to the set-trigger system of the needle guns.

Jäger rifles were also issued to the foot artillery and the Pioneers, and finally, displaced by more modern equipment, to German colonial troops. *Jägerbüchsen* M1871 were also issued to the *Kaiserliche Marine*, where they replaced 1854-pattern Dreyse needle guns. By 1905, nearly 7,000 examples of the 1871-type short rifle remained in the inventories of I. *Matrosen-Division* and the two dockyard divisions; by 1907, however, only the dockyard guns remained.

Jägerbüchsen M1871 were made by the Prussian government arsenal in Danzig, Gebrüder Mauser & Cie of Oberndorf am Neckar, and Waffenfabrik Steyr. They were identical mechanically to the standard M1871 and marked similarly, though 'J.G.MOD.71' was substituted for 'I.G.MOD.71' on the left side of the receiver. It can be difficult to distinguish between the Fraktur 'J' and 'I', but barrel length will resolve confusion.

The standard bayonet – the *Hirschfänger* M1871 – had a steel pommel, chequered leather grips, a recurved cross-guard, and a distinctive

11mm *Jägerbüchse Modell* 1871 (JB Mod. 71)	
Synonym:	Mauser short rifle M1871
Adoption date:	18 January 1876
Length:	1,240mm (48.82in)
	with bayonet
	1,733mm (68.23in)
	with Hirschfänger 71
Weight:	*without sling*
	4.475kg (9.87lb)
	with bayonet
	5.125kg (11.3lb)
	with Hirschfänger 71
Barrel length:	750mm (29.53in)
Chambering:	11 × 60mm, rimmed
Rifling type:	four-groove, concentric
Depth of grooves:	0.4mm (0.01in)
Width of grooves:	4.5mm (0.177in)
Pitch of rifling:	one turn in 550mm (21.65in), RH
Loading system:	single rounds
Front sight:	open barleycorn
Backsight:	a small leaf and a large leaf with a sliding extension, fixed to a standing block
Backsight setting:	*minimum* 300m *maximum* 1,600m
Muzzle velocity:	415m/sec (1,360ft/sec)
Bullet weight:	25g (386gn)

495mm swell-point sword blade. Some obsolescent *Hirschfänger* M1865 were bushed for the 1871-pattern rifles, for issue while M1871 bayonets were still in short supply, but these can be identified by three grip rivets instead of five.

The Pioneers were issued with the *Pionier-faschinenmesser* M1871, with a cast-brass hilt, a wrought iron or steel cross-guard, and a massive saw-backed blade measuring 48 × 3.7cm. The foot artillerymen had a bushed *Füsilier-Seitengewehr* M1860, the so-called *Artillerie- Seitengewehr* M1871, which had a pipe-backed blade. There was also a cleaning rod, sling, muzzle protector, sight cover, screwdriver, spare striker spring and a separate extractor tool.

11mm Karabiner Modell *1871*
(Kar. Mod. 71)

Experience in the Franco-Prussian War showed that carbines were useful to all cavalrymen – not just to dragoons and hussars – as well as to patrols, pickets, outposts, army postmen, and others who had been left to the mercies of French *franctireurs* during hostilities.

Developing a shortened version of the *Infanterie-Gewehr* M1871 was out of the question, as resources were concentrated on making new infantry rifles. A short-term solution was found by converting captured French Chassepot infantry rifles to carbine length, allowing *Aptierter Chassepot- Karabiner* to be adopted for the dragoons and hussars early in 1875. Issue was extended to the *Ulanen* (lancers) on 31 August 1876 and to the cuirassiers in November 1880.

A prototype Mauser carbine was eventually tested by the *Gewehr-Prüfungs-Kommission* in the spring of 1875. It was 1,000mm long and had the cylinder-and-octagon barrel form of the standard infantry rifle. There was also a cleaning rod, a sight

11mm *Karabiner Modell* **1871** (Kar. Mod. 71)	
Synonym:	Mauser carbine M1871
Adoption date:	31 August 1876
Length:	995mm (39.17in)
Weight:	3.425kg (7.55lb) without sling
Barrel length:	505mm (19.88in)
Chambering:	11 × 60mm, rimmed
Rifling type:	four-groove, concentric
Depth of grooves:	0.4mm (0.01in)
Width of grooves:	4.5mm (0.177in)
Pitch of rifling:	one turn in 550mm (21.65in), RH
Loading system:	single rounds
Front sight:	open barleycorn
Backsight:	a small leaf and a large leaf-and-slider attached to a standing block
Backsight setting:	*minimum* 200m *maximum* 1,200m
Muzzle velocity:	395m/sec (1295ft/sec)
Bullet weight:	25g (386gn)

The Karabiner *M1871, with the action closed.*

The Karabiner *M1871, top view.*

The Karabiner *M1871 with the bolt open.*

similar to the *Jägerbüchse* pattern (then experimental), and a bolt-handle which had been bent downwards toward the stock. However, the cleaning rod was soon discarded and a diminutive but otherwise conventional leaf sight was substituted for the full-size type. A reduced-charge cartridge minimized the blast and flash encountered when rifle ammunition was fired in such a short barrel.

The bolt mechanism of the *Karabiner* M1871 is almost identical with that of the 1871-type infantry rifle, apart from a turned-down operating handle. The cylindrical barrel tapers slightly towards the muzzle, and a leaf-type backsight lies immediately ahead of the octagonal section of the breech.

The one-piece walnut stock extends to the muzzle-cap, which has projecting 'ears' to prevent the front sight snagging in the carbine scabbard. A single band, held in place by a spring, carries the front sling swivel. The back swivel lies under the butt.

The Mauser carbine became the weapon of dragoons, hussars and lancers, except for NCOs and trumpeters (who retained single-shot pistols). Carbines were first issued to the cuirassiers in April 1884, but complete re-arming was delayed until 1888.

The *Karabiner* M1871 served until the approval of the *Karabiner* 88 in January 1890. Most had been discarded or handed to the *Landwehr* by the turn of the century, although many soldiers had taken the option of purchasing surplus examples for two Marks apiece. One thousand were sold to China in 1906 and A.L. Frank was still trying to sell 375 of them in 1911.

As the Prussian government arsenals had very little spare manufacturing capacity, the first 60,000 *Karabiner* M1871 were ordered from Österreichische Waffenfabriks-Gesellschaft in Steyr in the summer of 1875. They had all been delivered before the end of 1877.

Gebrüder Mauser & Cie made 3,000 *Karabiner* M1871 for Württemberg, and others for Prussia and Saxony. An order was also given to 'Productionsgemeinschaft Spangenberg & Sauer, Schilling &

Haenel' in Suhl, although these contractors – despite being considered a cartel – apparently fulfilled their obligations separately, as guns have been reported marked 'AGH' for Arbeitsgemeinschaft Haenel or 'V.C.S.' for Schilling.

A typical *Karabiner* M1871 bears its maker's mark along the top flat of the barrel octagon. The designation 'K.MOD.71', in Fraktur, lies on the left rear of the receiver. Serial numbers appear on the left side of the barrel, on the left side of the receiver, on the bolt head, and on the bolt-guide rib alongside the operating handle. A date appears on the right side of the receiver behind the bolt-handle, and a large crowned 'FW' or 'W' cipher may sometimes be seen struck into the left upper face of the barrel octagon.

Carbines were customarily accompanied by a leather backsight cover, a sling, a screwdriver, a spare mainspring and a cartridge-extracting tool. They were also often issued with a saddle-mounted scabbard called the *Karabiner-Futteral* 71, but did not accept bayonets.

1871-type Gendarmerie Carbines

Also known as the '*Landjäger-Gewehr*', the *Gendarmerie-Gewehr* M1871 was issued in Württemberg from 1876 onwards to the *Königliches Landjäger-Korps* (rural gendamerie). The guns were similar to the *Karabiner* M1871, but had a bayonet lug on the right side of the nose-cap and unit marks prefaced 'K.L.K.'.

Adopted on 29 November 1879, *Grenz-Aufseher-Gewehre* M1879 were made by Schilling of Suhl for the *Grenz- und Steueraufseher zu Fuss*, the Prussian corps of border guards and customs officials. Marked 'G.A.G.MOD.79' on the rear left side of the receiver, the guns were made by V.C. Schilling of Suhl in 1880–81 and bear the 'V.C.S.' mark on the barrel. They resemble the standard *Karabiner* M1871, but have two-position rocking 'L'-pattern backsights, Dreyse-style stock fittings and lug on the nose-cap for a sabre bayonet. *Gendarmerie* rifles are occasionally reported with the marks of C. G. Haenel of Suhl, but it is not known if they were contemporaneous with the Schilling examples.

MACHINE-GUNS

Front-Rank Patterns

7.9mm Maschinengewehr Modell *1908* (MG08)

This machine-gun was essentially similar to the earlier German Maxims, but improved in many details and mounted on the 1908-pattern sled mount or *Schlitten* 08. The mount no longer had wheels, nor was it specifically intended to be dragged – but it could be lifted and carried for short distances by two strong men.

A similar machine-gun was offered commercially by Deutsche Waffen- und Munitionsfabriken as the *Maschinengewehr Modell* 1909, which sold particularly well in Bulgaria, China, Romania, Turkey and elsewhere, but the company subsequently staked its hopes on the Parabellum (qv) to compete with the lightened Vickers guns.

Vickers, Sons & Maxim had already granted a production licence to Deutsche Waffen- und Munitionsfabriken of Berlin, which subsequently granted a sub-licence to the Prussian authorities. The

7.9mm *Maschinengewehr Modell* 1908 (MG08)	
Synonym:	German Maxim M1908
Adoption date:	1908
Length of gun:	1,095mm (43.11in) without booster
Weight of gun:	26.45g (58.3lb) with coolant
Weight of coolant:	4kg (8.8lb)
Weight of mounting:	32.5kg (71.65lb)
Barrel length:	720mm (28.35in)
Chambering:	7.9 × 57mm, rimless
Rifling type:	four-groove, concentric
Depth of grooves:	0.15mm (0.006in)
Width of grooves:	4.5mm (0.177in)
Pitch of rifling:	one turn in 240mm (9.45in), RH
Feed type:	100- or 250-rd fabric belt
Weight of loaded belt (250-rd type):	7.25kg (16lb)
Front sight:	open barleycorn
Backsight:	a pivoting bar and slide
Backsight setting:	*minimum* 400m *maximum* 2,000m
Rate of fire:	300 rd/min (without booster)
Muzzle velocity:	860m/sec (2,820ft/sec)
Bullet weight:	9.85g (152gn), 'S' type

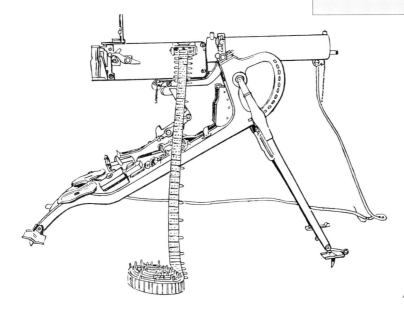

A drawing of the Maxim MG08.

A typical MG08 on its sled mount, with detachable water-jacket armour. Firepower International

Spandau rifle factory delivered the first government-made guns in 1910. The Maxims had the manufacturer's mark, designation and date under the serial number on top of the receiver, level with the feed block. Typical examples read 'M.G. 08.', 'D.W.M.', 'BERLIN' and '1911' or 'M.G.08.', 'Gwf.', 'SPANDAU' and '1912' in four lines. Information was repeated on the fusee-spring cover – for example, 'DEUTSCHE WAFFEN- UND' and 'MUNITIONSFABRIKEN' encircling 'BERLIN' and the date; or, alternatively, '✷ MASCH. GEW. 08. ✷ GEWEHRFABRIK' around 'SPANDAU' and the date. Other contractors were recruited during World War One (*see* Part Three: '1915').

Despite detail changes, however, the German MG08 remained quintessentially Maxim; the

A machine-gunner in a well-constructed defensive outpost in the Argonne observes the battlefield through a periscope. The MG08, mounted on a sled, has a water-cooled recoil booster and a prominent dent in the barrel casing. Markings on the ammunition boxes identify the machine-gun company of the 129th Infantry Regiment.

Drawings of the MG08, from a 1927-vintage handbook.

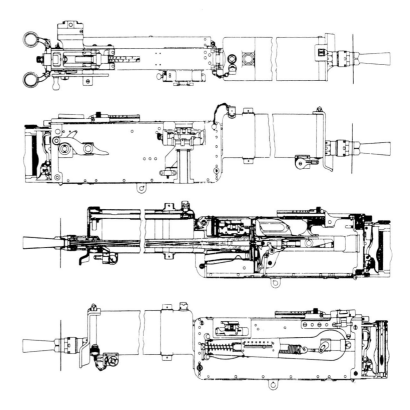

action is described in detail in the entry devoted to the Austrian Maxim-*Maschinengewehr Modell 1904* (M04) (qv). The standard mount was the *Schlitten* 08, though the *Dreifuss* 16 was introduced during World War One. Owing to its MG08/15-type mounting lugs, however, the 1916-pattern tripod mount required an adaptor. A special shielded mount was developed for use in fortifications; guns intended for naval service had pintle mounts; and a simplified auxiliary mount was produced during World War One for static use. Special anti-aircraft adaptors were also introduced.

Most MG08s were issued with covers and carrying slings. Two spare barrels were carried in a special container and a third either on the base plate of the sled mount or inside the back leg of the tripod. The *Rückstossverstärker S* (recoil booster) was introduced during World War One, together with a 'soaking container' to remove fouling.

The standard canvas ammunition belts could be carried in special *Patronenkasten* (PK) cartridge boxes, the PK11 holding two 250-round belts and the PK16 only one. Slings allowed the boxes to be carried when necessary. A few *aptierte Art* ('altered pattern') guns could handle a metal-link belt in addition to the canvas type.

The 100-round belt was often issued with a *Gurttrommel* (belt drum), accompanied by a carrying pouch, which could be attached to the gun once an auxiliary bracket had been added to the right side of the receiver beneath the feed way. Belt-cleaning brushes, pliers, and even a belt-filling machine known as the *Gurtfüller* 16 could be obtained.

The MG08 was also issued with a special optical sight, originally known as the *Zielfernrohr für Maschinengewehr*, which could be mounted on a special bracket offset on the rear left side of the receiver. Issued in a special protective leather case, most of these units were made by Zeiss or Goerz prior to 1914, but other contractors were subsequently recruited.

A steam hose and a water chest, customarily marked 'W' (*Wasser*), were issued to circulate coolant; and a water jacket heater could be fitted to prevent coolant freezing in cold climates. Canisters containing glycerine and oil were marked 'G' and 'O' respectively. Spare-parts boxes were marked 'E' (*Ersatzteile*), and repair kits were marked 'V'. Tools included an oil can, the 1908-pattern machine-gun wrench, and a clamp to allow pierced water jackets to serve until more permanent repairs could be made.

Armoured shields were provided to protect the water jacket from frontal or lateral hits, a heavy shield was available to protect the gun and mount, and armour-plated vests could even be issued to the gunners.

The MG08 was designed to be transported on a special handcart, with its own special parts box, or on a horse-drawn wagon accompanied by an ammunition limber. Pack equipment allowed the gun and sled to be transported by animals in hilly or mountainous districts.

7.9mm *Maschinengewehr Modell 1914 (MG14)*

The introduction of the light-pattern Vickers gun in 1911–12 threatened commercial sales of the Deutsche Waffen- und Munitionsfabriken-made Model 1909 Maxim. As a consequence, work began immediately to refine the Maxim under the supervision of an experienced engineer named Karl Heinemann.

Realizing that the size and weight of the Maxim receiver was due to the bulky toggle-lock, which folded downwards, Heinemann reached the same conclusion as the designers of the light-pattern Vickers gun and inverted the locking mechanism. Careful attention to the detail of individual parts enabled a much more compact feed unit to be incorporated, and the Parabellum is regarded as the lightest of all the true derivatives of the Maxim.

Extensive testing showed that the new gun was efficient enough to enter series production as the

---

7.9mm *Maschinengewehr Modell* 1914 (MG14)	
Synonym:	German Parabellum machine-gun
Adoption date:	1914
Length of gun:	1,225mm (48.23in)
Weight of gun:	10.15kg (22.4lb) without coolant
Weight of coolant:	1.5kg (3.3lb)
Weight of mounting:	not known
Barrel length:	705mm (27.75in)
Chambering:	7.9 × 57mm, rimless
Rifling type:	four-groove, concentric
Depth of grooves:	0.15mm (0.006in)
Width of grooves:	4.5mm (0.177in)
Pitch of rifling:	one turn in 240mm (9.45in), RH
Feed type:	100- or 250-rd fabric belt
Weight of loaded belt (250-rd type):	7.25kg (16lb)
Front sight:	open barleycorn
Backsight:	tangent-leaf type
Backsight setting:	*minimum* 400m *maximum* 2,000m
Rate of fire:	600–650rd/min
Muzzle velocity:	860m/sec (2820ft/sec)
Bullet weight:	9.85g (152gn), 'S' type

Schweres Maschinengewehr Modell 1913 (Model 1913 heavy machine-gun), virtually all Parabellums, even those made during World War One, being marked 'S.M. GEW. MOD. PARABELLUM 1913' and 'BERLIN' on the left side of the receiver. A decorative 'DWM' monogram lay on the receiver-top. Deutsche Waffen- und Munitionsfabriken's telegraphic code name 'Parabellum' had already achieved prominence on the Luger-designed pistol, but the angular style of the monogram used on the machine-gun differed greatly from the cursive pattern associated with the handgun.

World War One began before the Parabellum could be made in quantity, though a few had probably been delivered for trials with the German Army. Its later career is summarized in Part Three.

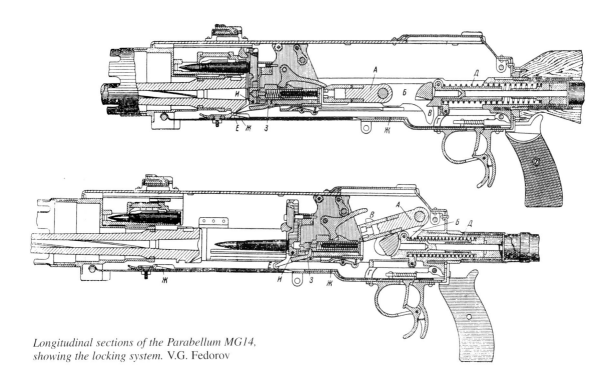

Longitudinal sections of the Parabellum MG14,
showing the locking system. V.G. Fedorov

7.9mm *Maschinengewehr Modell* 1914 (MG14)

Internal Arrangements

Although the operation of the Parabellum has a considerable affinity with the Maxim, the internal layout is radically different. The return spring lies directly behind the bolt, projecting into a tube in the shoulder stock, and the toggle-breaking system has changed. Two lugs on the base of the crank strike abutments in the base of the frame to break the locking joint upwards instead of downwards. Radical alterations have been made to the feed system – extremely wasteful of space in the Maxim – to allow the Parabellum receiver to be much shallower. Cyclic rate is increased by camming the barrel forward before the lock reaches the limit of its rearward travel. The return spring cannot be used to adjust the fire-rate, as in the Maxim, but this is a small price to pay for the savings in weight.

External Appearance

The Parabellum has a squared receiver and a wooden shoulder stock or butt. The charging slide lies on the right side. A trigger-guard with a distinctive spur, which contains a pivoting spring-loaded trigger-blocking safety lever, lies under the receiver immediately ahead of the pistol-grip. Guns used aboard Zeppelin airships often had conical flash-hiders.

A tangent-leaf backsight lies on top of the receiver, hinged at the back of its bed. The front sight is customarily a small ring, although a multi-bar 'gate' pattern may be fitted to observers' guns. Water-cooled guns have a mounting eye on a strap encircling the rear of the water jacket, but the air-cooled versions normally have a pintle mount beneath the receiver in line with the feed aperture.

Second-Rank Patterns

8mm Maxims

The older MG01 Maxims (*see* Part One) were restricted to training and fortification use after the MG 08 became available in quantity, though survivors remained in store when World War One began. Some of the original *Kaiserliche Marine* guns were also still in service in 1914, shipboard guns having been issued with tripods instead of Army-style sleds. Photographs taken of the *Emden* machine-gun crews defending Direction Island in November 1914 clearly show Deutsche Waffen- und Munitionsfabriken-made examples of the MG07 (the commercial designation for an improved form of the MG01) on tripod mounts.

A gunner from SMS Emden *poses with one of the cruiser's four Maxim machine-guns, mounted on a DWM-pattern tripod mount. Direction Island, 9 November 1914.* Imperial War Museum

(Below) *The MG01 and* Schlitten *03 with their cart and limber.*

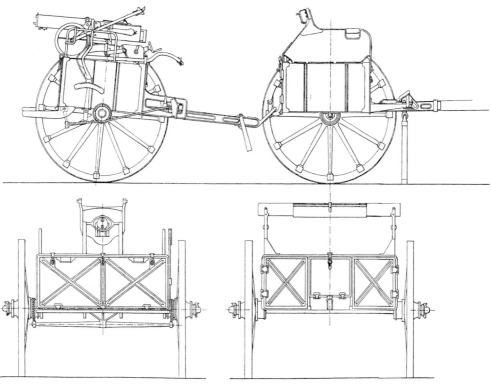

8　Turkey

Turkish guns are notoriously difficult to classify, as local alterations often blurred their origins even though basic weapons customarily follow specific patterns – for example, Mauser or Peabody-Martini. Wholesale changes were made to ex-German Gew. 88 by the Ankara ordnance factory in the period between World Wars One and Two, and a multitude of non-standard bayonets will also be found.

HANDGUNS

7.63mm Mauser-Selbstladepistole C96

One thousand ten-shot C96 pistols were ordered from Mauser in July 1897 accompanied by one thousand stocks, one thousand cleaning rods and 250,000 cartridges. These were duly delivered to Constantinople in the summer of 1898. They were identical to the pistols being tested in Germany (qv), but had 50–1,000m tangent-leaf sights graduated in Arabic numerals. The date '1314', 1897 appears beneath a *Toughra* on the frame-panel directly above the left grip.

9mm FN-Browning Model 1903

About 10,000 of these guns were acquired for the paramilitary police service in 1906–07. The guns display the encircled *Toughra* of Sultan Abdülhamid II on the top of the slide, together with a property mark and an issue number in Arabic – for example, '6869' – on the right side of the slide halfway between the trigger and the muzzle. The guns are not dated, but holsters have been seen marked for '1324' (1907).

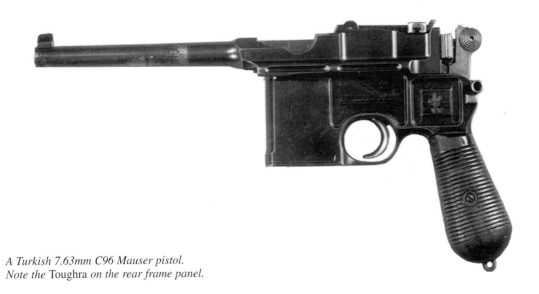

A Turkish 7.63mm C96 Mauser pistol.
Note the Toughra *on the rear frame panel.*

RIFLES

Front-Rank Patterns

7.65mm Infantry Rifle Model 1903

Experimental Gew. 98-type rifles were tested extensively in 1902, allowing a 200,000-gun order to be placed with Waffenfabrik Mauser in an era when ninety-six field guns were acquired from Krupp. Deliveries totalling 207,700 were made from Oberndorf in 1903–08.

The M1903 resembled the contemporary German service rifle externally, but it had a conventional tangent-leaf backsight, a hand guard extending from the receiver ring to the band, and a bayonet lug under the simple nose-cap. The cocking-piece was longer and heavier than its German

equivalent. The stock had a pistol-grip, and swivels lay under the band and butt.

Spitzer or pointed-bullet ammunition, introduced in 1910, led to a change in sights, but the unfamiliarity of Arabic notation can make this difficult to determine.

7.65mm Cavalry Carbine Model 1908

7.65mm Cavalry Carbine Model 1908	
Synonyms:	Turkish Mauser carbine M1905 or M1908
Adoption date:	1908
Length:	1,045mm (41.14in)
Weight:	3.75kg (8.27lb)
	without sling
Barrel length:	550mm (21.65in)
Chambering:	7.65 × 53mm, rimless
Rifling type:	four-groove, concentric
Depth of grooves:	0.13mm (0.005in)
Width of grooves:	4.2mm (0.165in)
Pitch of rifling:	one turn in 250mm (9.84in), RH
Magazine type:	staggered-row internal box
Magazine capacity:	5 rounds
Loading system:	charger or single rounds
Cut-off system:	none
Front sight:	open barleycorn
Backsight:	tangent-leaf type
Backsight setting:	*minimum* 200m *maximum* 1,600m(?)
Muzzle velocity:	615m/sec (2,020ft/sec)
Bullet weight:	13.65g (211gn)

7.65mm Infantry Rifle Model 1903	
Synonym:	Turkish Mauser M1903
Adoption date:	December 1902(?)
Length:	1,242mm (48.9in) *with bayonet* 1,765mm (69.5in)
Weight:	*without sling* 4.175kg (9.2lb) *with bayonet* 4.575kg (10.09lb)
Barrel length:	740mm (29.13in)
Chambering:	7.65 × 53mm, rimless
Rifling type:	four-groove, concentric
Depth of grooves:	0.13mm (0.005in)
Width of grooves:	4.2mm (0.165in)
Pitch of rifling:	one turn in 250mm (9.84in), RH
Magazine type:	staggered-row internal box
Magazine capacity:	5 rounds
Loading system:	charger or single rounds
Cut-off system:	none
Front sight:	open barleycorn
Backsight:	tangent-leaf type
Backsight setting:	*minimum* 400m *maximum* 2,000m
Muzzle velocity:	650m/sec (2,130ft/sec)
Bullet weight:	13.65g (211gn)

This gun, virtually a short rifle, was acquired to replace the earlier 1890 pattern in the hands of the cavalry and artillerymen. Customarily known as the 'M1905', a Mauser factory photograph dated 18 January 1909 is labelled 'M. 08 Karabiner, Türk. Visier.', and there seems little doubt that this is the correct designation.

The guns have a pistol-grip stock, a spring-retained barrel band, and a simple nose-cap carried upwards to protect the front-sight blade. The bolt-handle turns down towards the stock; sling mounts lie on the left side of the band and butt.

Second-Rank Patterns

7.65mm Infantry Rifle Model 1893

The approval of the Spanish Mo. 1893 led Turkey to request an improvement on their 1890-pattern rifle. Samples were sent to Turkey early in September 1893, allowing an order for 200,000 of a modified pattern to be placed shortly before Christmas that same year. Deliveries commenced on 30 July 1894, and the last of 201,900 guns left Oberndorf on 21 September 1896. They included 1,800 for the Sultan's guard, with polished stocks and nickelled metalwork. It is believed that another order – for 100,000 guns – was placed in the spring of 1897, when friction between Greece and Turkey threatened to escalate into war. Deliveries were still being made in 1899.

The 1893-pattern Turkish rifle duplicated the Spanish Mo. 1893, except for its conventional cylindrical bolt and a unique cut-off lever on the right side of the receiver beneath the bolt way. An extension of the bolt-stop doubled as the left Zcharger guide, and Arabic markings lay on the receiver.

7.65mm Infantry Rifle Model 1890

Tests begun in Germany in the late summer of 1889 allowed an experimental 7.65mm charger-loading Mauser rifle to go to Turkey with Colmar,

7.65mm Infantry Rifle Model 1893	
Synonym:	Turkish Mauser M1893
Adoption date:	1893
Length:	1,235mm (48.62in)
	with bayonet
	1,696mm (66.77in)
Weight:	*without sling*
	4.110kg (9.06lb)
	with bayonet
	4.760kg (10.5lb)
Barrel length:	740mm (29.13in)
Chambering:	7.65 × 53mm, rimless
Rifling type:	four-groove, concentric
Depth of grooves:	0.13mm (0.005in)
Width of grooves:	4.2mm (0.165in)
Pitch of rifling:	one turn in 250mm
	(9.84in), RH
Magazine type:	protruding single-row box
Magazine capacity:	5 rounds
Loading system:	charger or single rounds
Cut-off system:	on receiver alongside
	bolt way
Front sight:	open barleycorn
Backsight:	tangent-leaf type
Backsight setting:	*minimum* 250m
	maximum 2,000m
Muzzle velocity:	630m/sec (2,065ft/sec)
Bullet weight:	13.65g (211gn)

7.65mm Infantry Rifle Model 1890	
Synonym:	Turkish Mauser M1890
Adoption date:	6 August 1890(?)
Length:	1,237mm (48.7in)
	with bayonet
	1,698mm (66.85in)
Weight:	*without sling*
	4.015kg (8.85lb)
	with bayonet
	4.695kg (10.35lb)
Barrel length:	740mm (29.13in)
Chambering:	7.65 × 53mm, rimless
Rifling type:	four-groove, concentric
Depth of grooves:	0.13mm (0.005in)
Width of grooves:	4.2mm (0.165in)
Pitch of rifling:	one turn in 250mm
	(9.84in), RH
Magazine type:	protruding single-row box
Magazine capacity:	5 rounds
Loading system:	charger or single rounds
Cut-off system:	none
Front sight:	open barleycorn
Backsight:	leaf and slider type
Backsight setting:	*Minimum* 250m
	maximum 2,000m
Muzzle velocity:	630m/sec (2,065ft/sec)
Bullet weight:	13.65g (211gn)

Freiherr von der Goltz. Trials with two improved rifles were successfully undertaken in June 1890, and, on 20 July 1890, the testers recommended that the M1890 should be adopted immediately. Turkey promptly invoked the substitution clause in the original contract, and the remainder of the 1887 contract – 280,000 rifles and 46,000 carbines – was to be completed with 1890-pattern guns.

The 7.65mm cartridge had a rimless bottle-necked brass case containing 2.65g of Rottweil nitro-cellulose propellant; loaded rounds were about 78mm overall and weighed 27.2g. The M1890 had a plain barrel, stepped to allow for expansion during rapid fire without upsetting the stock bedding. A one-piece sear was fitted and the magazine could be removed by pressing a catch set into the front face of the trigger-guard. The combination of the left charger guide with the bolt-release catch was patented in Germany in June 1889; this allowed the left charger guide to fit chargers of slightly varying width. However, though seemingly advantageous, the gradual weakening of the release-catch spring was a drawback, and so, apart from the M1893, the experiment was not repeated.

A short hand guard ran forward from the sight base, though it did not reach the solitary sprung band. The nose-cap assembly was simpler than the 1887 type, carrying a bayonet lug on the underside, whilst sling swivels lay beneath the band and butt. Graduated in Arabic numerals, the backsight was carried on a sleeve around the barrel.

The first M1890 rifles were assembled in Oberndorf in July 1891. Production was speedy, as the 200,000th example was completed in April 1893 and the last consignment left for Turkey in December. A few cavalry carbines were made experimentally in the spring of 1893, but Sultan Abdülhamid II ordered the adoption of something other than another Mauser and the Krag-Jørgensen was selected in September 1893. Only trial batches were ever delivered by Österreichische Waffenfabriks-Gesellschaft before the project was abandoned.

7.65mm Peabody-Martini conversions

Large numbers of single-shot Peabody-Martini rifles were converted for the 7.65 × 53mm cartridge in 1911–12, shortly before the First Balkan War began. These guns had new short barrels, fore-ends, hand guards and sights.

Substantial quantities of Martini-Henry rifles acquired from Britain in the late nineteenth century were also altered. Owing to the changes in barrels, stocks and sights, however, it can be difficult to tell the US- and British-made guns apart.

Obsolete Patterns

9.5mm Infantry Rifle Model 1887

Trials began in 1886 to find a magazine rifle to replace the Peabody-Martini, and an order for

9.5mm Infantry Rifle Model 1887	
Synonym:	Turkish Mauser M1887
Adoption date:	1887
Length:	1,250mm (49.21in)
	with bayonet
	1,720mm (67.7in)
Weight:	*without sling*
	4.245kg (9.36lb)
	with bayonet
	4.965kg (10.95lb)
Barrel length:	772mm (30.4in)
Chambering:	9.5 × 60mm, rimmed
Rifling type:	four-groove, concentric
Depth of grooves:	0.15mm (0.006in)
Width of grooves:	3.53mm (0.139in)
Pitch of rifling:	one turn in 500mm
	(19.68in), RH
Magazine type:	tube in fore-end
Magazine capacity:	8 rounds
Loading system:	single rounds
Cut-off system:	yes
Front sight:	open barleycorn
Backsight:	tangent-leaf type
Backsight setting:	*minimum* 200m
	maximum 1,600m
Muzzle velocity:	535m/sec (1,755ft/sec)
Bullet weight:	18.4g (284gn)

The Turkish M1887 Mauser short rifle or carbine. Hans-Bert Lockhoven

500,000 rifles and 50,000 carbines was ultimately shared by Waffenfabrik Mauser and Ludwig Loewe & Cie. Before work began, however, the contractors decided that Mauser should make all the guns for Turkey whilst Loewe accepted contracts for the *Reichsgewehre*.

The M1887 rifle chambered a powerful black-powder cartridge created by necking the *Reichspatrone* to accept a small-diameter bullet. Assembled rounds were 75.4mm long, weighed 36g, and contained 4.55g of black powder. Additional power required an additional locking lug, but the rifle was otherwise similar to the Gew. 71/84. It had a straight-wrist stock, a single spring-retained barrel band, and a nose-cap with a bayonet lug on the right side. Swivels lay on the underside of the nose-cap and in the front web of the trigger-guard, and a German-type backsight was attached to the barrel ahead of the octagonal chamber reinforce.

The first consignment of 1,350 M1887 rifles left Oberndorf on 30 May 1888, but in 1890, owing to progress with the 1889-type Mauser in Belgium, the Turkish authorities invoked the substitution clause in the original contract. Production of the M1887 stopped on 24 December 1890, after about 220,000 rifles and only 4,000 carbines had been made. The last guns were dispatched to Turkey at the beginning of March 1891.

11.43mm (0.45in) Infantry Rifle Model 1874
Turkey was among the first to buy large quantities of Peabody-Martini rifles, ordering 400,000 in 1874. These are difficult to distinguish from the British Martini-Henry, but have a safety lever ahead of the trigger-guard. Most of the rifles were fully stocked, had two barrel bands, and accepted

socket bayonets; others, possibly for riflemen and élite units, were issued with leather-gripped sabre bayonets attaching to a lug on the right side of the muzzle. Sling swivels lay under the front barrel band, immediately ahead of the trigger-guard, and on the under-edge of the butt. A large cocking indicator pivoted on the right rear side of the receiver.

11.43mm (0.45in) Infantry Rifle Model 1874

Synonym:	Turkish Peabody-Martini M1874
Adoption date:	1874
Length:	1,245mm (49.02in) *with bayonet* 1,755mm (69.1in) with socket type
Weight:	*without sling* 4.37kg (9.63lb) *with socket bayonet* 4.785kg (10.55lb) with socket type
Barrel length:	845mm (33.27in)
Chambering:	11.43 × 58mm, rimmed
Rifling type:	five-groove, concentric
Depth of grooves:	0.2mm (0.008in)
Width of grooves:	not known
Pitch of rifling:	one turn in 560mm (22.05in), RH
Loading system:	single shots
Front sight:	open barleycorn
Backsight:	leaf-and-slider type
Backsight setting:	*minimum* 100yd *maximum* 1,400yd
Muzzle velocity:	420m/sec (1,380ft/sec)
Bullet weight:	31.1g (480gn)

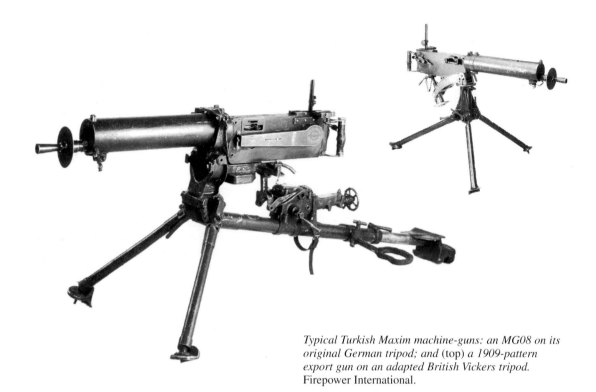

Typical Turkish Maxim machine-guns: an MG08 on its original German tripod; and (top) *a 1909-pattern export gun on an adapted British Vickers tripod.* Firepower International.

11.43mm (0.45in) Cavalry Carbine Model 1874

Turkey ordered 25,000 Peabody-Martini carbines for its cavalry, placing further (smaller) orders into the late 1870s until total acquisitions approached 50,000. The guns had half-length fore-ends held by a single band, but lacked bayonets.

11mm (0.433in) Remington Rifle

Surprisingly, quantities of these Rolling Block guns remained in store in 1914. Most were apparently retrieved from troops serving in Egypt, part of the Ottoman Empire until occupied by Britain in 1882. The rifles had three sprung bands and a bayonet lug on the right side of the muzzle. They were originally ordered in 1867 from E. Remington & Sons of Ilion, New York, but some have been reported with the marks of Em. & L. Nagant of Liége, Remington's licensee, and it is suspected that orders were still being fulfilled in

the mid-1870s. The guns were typically 1,280mm long, had 890mm barrels rifled with a five-groove right-hand twist, and weighed about 4,150g. Their backsights were graduated to 1,000yd.

MACHINE-GUNS

The Turkish forces were largely equipped with Maxims, although a few of the Gatlings that had been acquired in the 1870s and 1880s remained in service in 1914. Most of the Maxims seem to have been 7.65mm MG09 patterns supplied by Deutsche Waffen- und Munitionsfabriken, though German-service 7.9mm MG08 were supplied during World War One.

The guns were usually distinguished by the *Toughra* mark, prior to 1908, or by a star-and-crescent symbol. They bear maker's marks, serial numbers and backsight graduations in Arabic.

PART THREE: WORLD WAR ONE

The 'War to end Wars' had its origins in the alliances, treaties and understandings that forced countries, often grudgingly, to assist each other in times of distress.

Space can be spared here only for the barest outline of World War One, but there are many excellent sources of detailed information. One of the best – and certainly among the most readable – is Captain Sir Basil Liddell-Hart's classic *History of the First World War* (Cassell & Co., 1970),

which reduces complex tactics and strategy to pithy, readily understood summary.

A state visit to Sarajevo, capital of Bosnia–Herzegovina, ended on 28 June 1914 with the assassination of the heir apparent to the Habsburg Empire and his consort. The deaths of Archduke Franz Ferdinand and Sophie von Hohenberg lit the touch-paper of the European powder keg by drawing attention to the demands of the Pan-Slav movement, but the keg had been firmly put in place by generations of suspicion

Men of Landesaustellung Windhoek, *the colonial gendarmerie in German South-West Africa, pose for the camera shortly before World War One. They were armed – most unusually – with the 7.65mm Roth-Sauer pistol.* James Hellyer

and mistrust. Much of Europe had achieved an uneasy tolerance by 1914 – Admiral Sir George Warrender, leaving Kiel in midsummer with the British fleet signalled his German hosts 'Friends now, friends forever' – but the second of two wars had only just finished in the Balkans, where mutual hatred undermined newly-won stability.

Austria-Hungary used the Sarajevo incident to pressurize Serbia; a state of war between the two was declared on 28 July 1914, but Russia sprang to Serbia's defence. Muscles flexed throughout Europe as the Triple Alliance – Germany, Austria-Hungary and Italy – squared up to the Triple Entente of Britain, France and Russia.

Within a week, intent on humiliating France for a second time within fifty years, Germany had invaded Belgium and forced Britain to defend the honour of the Triple Entente. Europe, precipitately, was at war.

The Germans charged through Belgium, intent on overrunning northern France and encircling Paris to bring the French government to its knees. However,

tinkering by General Helmuth von Moltke (Germany's chief of staff) had compromised the Schlieffen Plan, keystone of German hopes, and the invaders were abruptly halted by battles on the Marne and Aisne rivers that foretold horrors to come.

Conflict in the West thereafter stabilized into a war of attrition, as the front-lines flexed back and forth with each thrust and counter-thrust. The optimism of 1914 gave way to new cynicism as the battles of the Somme, Verdun and Ypres changed little but the toll of the dead and the maimed.

The Eastern Front, with its marshes and endless plains, was characterized by much more fluid warfare. German and Austro-Hungarian forces rebuffed an initial Russian invasion of Poland and Galicia in 1914, successfully mounting a counter-offensive in the spring of 1915, but in 1916 the Russians regained much of the territory that had been lost. However, the human cost of the Brusilov offensives, disintegration of morale and, ultimately, the October Revolution in 1917 ended Russian resolve. The Treaty of Brest-Litovsk, signed on 3 March

Sailors of the landing party of the cruiser SMS Gneisenau *muster ashore, probably in Tsingtau in 1914. They were issued with Mauser rifles, Parabellum pistols and machine-guns from the shipboard armoury.*

Recruits and their NCO instructor pose for the camera after their basic training. The rifles – all Gewehre 98 *– apparently have a distinguishing band (paint?) around the fore-end, possibly indicating suitability only for drill purposes.*

1918, gave the Central Powers far more territory by negotiation than they had achieved by fighting.

In the south, the ill-fated seaborne invasion of Gallipoli and the Dardanelles by the Allies in 1915 distracted Turkish forces (at terrible cost) from the invasion by Russia of Persia and the Turkish Caucasus. In Mesopotamia and Egypt, fated to make little progress until the end of the war, British units struggled to overcome not only Turkish resistance but also the horrors of starvation and disease. Along the valley of the Isonzo river, the Austro-Hungarians and the Italians fought intermittently and inconclusively to a bloody standstill.

The exploits of raiders and cruiser squadrons at sea in 1914, greatly though they captured public imagination, achieved little of strategic significance. Long-standing traditions of warfare soon gave way to a grim maritime siege in which an untried child of technology – the German submarine – sought to strangle British maritime trade, unfettered by notions of chivalry. The only major fleet action to be fought during World War One, the Battle of Jutland (31 May–1 June 1916), petered out to an inglorious draw which showed that advances in warship design had far outstripped the ability to communicate at sea.

The entry of the USA into the war in April 1917 proved decisive. Even though the collapse of Russia allowed new divisions to be thrown into the fighting on the Western Front, the Spring Offensive of 1918 achieved no more for the Germans than the Somme and Verdun had done for the Allies two years earlier. Small gains bought at vast cost in lives and munitions were reversed with the assistance of the American Expeditionary Force, and the Allies had finally gained an ascendancy – even retaking parts of Belgium – when the empires of Europe fell apart in the autumn of 1918.

9   1914

AUSTRIA–HUNGARY

Organization

The men of the *kaiserlich und königlich Armée* – known in Hungary as *Közös haderö* – formed the regular forces. However, the principal components of the Austro-Hungarian Empire each had their own second-line establishment, known in Austria as the *kaiserlich-königlich Landwehr* (*k.k. Landwehr*) and in Hungary as the *Magyar kiralyi honvédség* (*Honvéd*).

Backing these was the *Einsatzreserve* (Austria) or *Pottartalék* (Hungary), and the third-rank forces of the *k.k. Landsturm* and the *Magyar kiralyi népfölkelés*. The Dual Monarchy was divided into sixteen army corps districts, subdivided into 112 recruiting areas.

The liability to serve extended from the year of a man's nineteenth birthday to the year of his forty-second. The first stages were decided by ballot, one group serving two or three years in the regular Army, another serving the *Landwehr* or *Honvéd*, and a third forming the *Einsatzreserve* or *Pottartalék*.

Regulars spent nine or ten years in the reserve when their front-line duties were completed and were then transferred briefly to the first-draft *Landsturm*; *Landwehr* personnel, reservists and first-draft *Landwehr* personnel were transferred to the second-draft *Landsturm* at the age of thirty-three.

Issue of Weapons

On 31 July 1914, the Austro-Hungarian armed forces could muster 118,000 M67/77, M73/77 and M77 Werndl rifles, carbines and *Extra-Corps-Gewehre* chambered for the 11mm M77 or M82 cartridges. There were also about 1.3 million M86/90 and M88/90 rifles – plus 80,000 M90 carbines and short rifles – chambered for the 8mm M88 or M90 cartridges; and about 850,000 M95 rifles, carbines and *Stutzen* chambering the 8mm M93 cartridge.

These totals were scarcely enough to satisfy immediate needs, even though monthly production capacity was reckoned to be about 75,000 rifles and 15,000 carbines, and so the Steyr (Österreichische Waffenfabriks-Gesellschaft) and Budapest (Fegyver és Gépgyár) factories were put on a war footing.

There were also 2,761 machine-guns on the inventory, which may have been more than served the German armies at that time. A machine-gun company attached to each infantry regiment had four sections of two guns apiece in August 1914. An officer with the rank of *Hauptmann* commanded four junior officers, sixty-four gunners in eight-man teams, nine reserves and forty-six pack-animal drivers; twenty-four mules, sixteen horses and seven wagons completed the establishment.

No. 1 mule carried the gun, the tripod and 500 rounds of ammunition; No. 2 carried 2,000 rounds; No. 3 had the armoured shield; Nos. 4 and No. 5 carried ammunition; and No. 6 carried ammunition and a tool kit.

The establishment of mountain machine-gun units was considerably lower than the infantry formations – merely three officers, eighty-two men and forty-four pack animals – but the status of the machine-gunners was soon disturbed by the war.

Sent in May 1917 by a member of a munitions column attached to Kaiserjäger *Regiment Nr 29, this shows a group of soldiers armed with 1895-type Mannlichers.*

7mm Repetiergewehr *M14*

When mobilization for war began in earnest in the summer of 1914, the Austro-Hungarian authorities were woefully short of modern rifles. Although this shortage was temporarily alleviated with German help, possibly including large numbers of Werndls from dealers' stores, another source of supply was found in the storerooms of Österreichische Waffenfabriks-Gesellschaft.

Ultimately, therefore, the government of the Dual Monarchy received nearly 67,000 Mauser-system rifles being made for the government of Mexico under the designation 'Mo. 1912'. The sling swivels were altered to conform to standard Austro-Hungarian practice; the guns were otherwise similar to the German *Gewehr* 98 (qv), except for the tangent backsight and the design of the nose-cap. They bore the Mexican eagle-and-cactus mark above the chamber and Austrian acceptance marks.

Twelve million 7 × 57mm rimless cartridges were ordered from Hirtenberg Patronenfabrik in the autumn of 1914, loaded with bullets weighing about 9g. Cartridge boxes, which held three loaded five-round chargers, were usually marked '15 stuck 7mm M14 Mauserpatrone' in black on a green label.

6.5mm Repetiergewehr *M3/14*

Shortly after the Second Balkan War had ended in 1913, Österreichische Waffenfabriks-Gesellschaft accepted an order from Greece for 200,000 Mannlicher-Schönauer 6.5mm rifles. Though many of these guns had already been delivered when World War One began, work to complete the order was still underway in Steyr.

Hirtenberg Patronenfabrik and the Roth company were recruited to make 6.5 × 54mm rimless cartridges, and the first 5,000 M3/14 guns (altered to accept Austrian-style slings) were issued to the Polish Legion in November 1914. Ammunition cartons were marked '15 St. 6,5mm griechische scharfe Gew.-Patrone'.

Surviving Mannlicher-Schönauer guns and parts were given to Italy after 1918. They were subsequently refurbished by Ernesto Breda, and, ironically, shipped to Greece.

GERMANY

Organization

When World War One began, the Imperial Army was still an amalgam of the four principal state

This photograph, allegedly taken in action in 1914, more probably shows manoeuvres: the machine-guns are too exposed and there are too many men in the trench to be convincing – and the puffs of smoke have clearly been added.

forces: Prussia, Bavaria, Saxony and Württemberg. The Bavarian and Saxon armies retained some independence, promoting officers on separate Army Lists, but the Prussians and the Württembergers were integrated in a single system.

Germany was divided into twenty-four military districts in 1914, including three in Bavaria; each district not only contributed an army corps, but also provided depots and recruiting facilities for individual units.

Liability to serve began on a boy's seventeenth birthday and lasted until the man was finally discharged from his obligations after reaching the age of forty-five. Service was initially with the *Landwehr* I. *Aufgebot* ('first-draft') until, at the age of twenty, a recruit was called up for examination. This resulted in a posting to the regular army for two or three years, depending on the branch of the service; to the Supplementary (*Ersatz*) Reserve for a period of twelve years; or rejection as unfit for military life.

Service in the regular army was followed by four or five years in the Reserve, and then eleven years in *Landwehr* I. *Aufgebot* before transfer to the second-rank home defence force, the *Landsturm* II. *Aufgebot*, at the age of thirty-nine. Those that had served in the Reserve since their initial call-up were released at the age of thirty-two,

trained men going to the second-draft *Landwehr* for seven years and untrained men to the first-draft *Landsturm* for the same period. Everyone was eventually transferred to the *Landsturm* II. *Aufgebot* in the year of their thirty-ninth birthday.

The distinctions between Reserve and *Landwehr* units were rapidly eroded after 1914, as units were absorbed into the regular army and 'war formations' were raised in haste.

Issue of Weapons

The standard long arms when mobilization began were the *Gewehr* 98 and *Karabiner* 98 AZ. The standard handgun was the *Pistole* 08, though, according to the 1914 edition of *Bewaffnung des Deutschen Heeres*, most of the gunners, drivers and non-commissioned officers of the field artillery still carried old M79 and M83 revolvers.

Production was still meeting requirements. Sufficient Mauser rifles had been delivered to equip the regular troops and a substantial proportion of the Reserve, and converted charger-loading *Gewehre* 88/05 had answered the shortfall. Deutsche Waffen- und Munitionsfabriken and Erfurt had delivered about 250,000 *Pistolen* 08 between them, though the figure is difficult to assess accurately.

Most of the men and junior NCOs of the infantry-men, *Jäger und Schützen* (riflemen), *Maschinen-gewehr-Abteilungen* (independent machine-gun units) and Pioneers carried the *Gewehr* 98 and the long S98 or S98/05 (Pioneers only); exceptions included the cyclists, who carried the Kar. 98 AZ and S84/98, and the men of the machine-gun companies, who carried pistols.

The junior NCOs and men of the dragoons, hussars, *Ulanen* (lancers), *Jäger zu Pferde* (mounted riflemen), Bavarian *schwere Reiter* (heavy cavalry) and *Chevaulegers* (light horsemen), and Saxon carabineers carried *Karabiner* 98 AZ. Bayonets were not issued until November 1914.

Field artillerymen carried pistols, but *Karabiner* 98 AZ with S98/05 were issued to the junior NCOs and other ranks of the foot artillery regiments. The NCOs and men of the Train battalions had *Karabiner* 98 AZ, issued prior to 1914 without bayonets.

The guns carried by men of the *Verkehrstruppen* (transport and lines-of-communication troops) varied according to unit. Junior NCOs and men of the *Eisenbahntruppen* (railway troops) carried *Gewehre* 98 and S98/05; senior NCOs had pistols. Junior NCOs, Pioneers and cyclists of the *Kraftfahrtruppen* (transport troops) carried *Karabiner* 98 AZ and S84/98. Most of the junior NCOs and men of the *Telegraphentruppen* (telegraph troops)

and the *Luftschiffertruppen* (airship units) were armed with *Karabiner* 98 AZ and KS98.

As far as the *Kaiserliche Marine* was concerned, the advent of the special Navy Parabellum, the *Pistole* 1904, coincided with an acceleration of German naval construction. Plans laid before the *Reichstag* in September 1905 had clearly been overtaken by events in Britain, and the Navy Minister, Admiral Alfred von Tirpitz approved a revised scheme on 4 June 1909.

Each German warship carried substantial quantities of small arms for duties ranging from the destruction of floating mines to arming small boats or landing parties. The battleship SMS *Friedrich der Grosse*, fleet flagship at Jutland, carried 385 *Gewehre* 98 and nearly 100 *Pistolen* 04 (excluding the extra guns carried by the admiral's staff), and the battle-cruiser *Blücher* – lost during the Battle of the Dogger Bank on 24 January 1915 – carried 280 rifles and 100 pistols.

Though SMS *Blücher* had been the artillery test ship in 1914 – carrying more small arms than normal – cruisers were designed for colonial service, sea-lane protection or raiding, and always had unusually high quotas of small arms. SMS *Emden* carried seventy *Gewehre* 98 and forty-six Parabellums on her ill-fated cruise in 1914 plus an *Auslandszuschlag* ('overseas supplement') of fifty

Sailors of the battleship Kaiser Wilhelm der Grosse *pose for the camera in January 1912. The rifles are* Gewehre 98.

*The pre-dreadnought battleship
Kaiser Wilhelm der Grosse
carried more than 200 rifles in
the shipboard armoury.*

rifles and five pistols. Even a fleet collier carried three obsolete rifles and a handgun (customarily a *Reichsrevolver*); the cadet-ship *König Wilhelm* had 900 rifles and seventy handguns!

Once World War One began, however, the *Kaiserliche Marine* authorities had to supply not only the ships of the battle-fleets but also the land-based *Marinekorps* securing the coastward flank of the Western Front in Flanders. Serviceable small arms were soon in very short supply.

It has been claimed that the Germans had more than 50,000 machine-guns ready for service in 1914, but the true figures are believed to have been about 1,600 in the Army and 400 in the Navy. Cruisers and battleships carried several Maxim machine-guns (usually four), which could be mounted on the ships' boats for short-range defence or taken ashore on a light tripod mount.

9mm Pistole 04

By 1911, Deutsche Waffen- und Munitionsfabriken was claiming to have made about 20,000 Parabellums for the German Navy – though probably only by including 'on order' guns which were

*French prisoners are mustered
into captivity in Thuringia early
in 1915. The guard on the left has
a Gewehr 88/05 in his hand and
another slung over his left shoulder.*

not delivered until shortly after World War One began. By October 1914, however, the Kiel dockyard had only six Parabellums in store.

The Navy-office weapons department noted on 6 November that Deutsche Waffen- und Munitionsfabriken was to supply 1,500 pistols immediately, 500 in January 1915, 1,000 in February, a similar quantity in March, and 1,500 in April, suggesting that a new order had been placed in the autumn of 1914. Nearly 6,000 *Pistolen* 04 had been delivered to Kiel by August 1916, when 'orders were complete' and another contract was given to Deutsche Waffen- und Munitionsfabriken.

7.9mm Gewehr 98

Combat experience showed that the Lange-system backsight of the Gew. 98 was inefficient; the guns shot very high at ranges of 50m or 100m, as the minimum sight setting was 400m. Fighting was taking place at much closer ranges than had been envisaged, and so a *Hilfskorn* (auxiliary sight) was added behind the front sight so that a Lange sight set for 400m was regulated for 100m.

7.9mm Parabellum-Maschinengewehr 14

When fighting began, procurement of the Parabellum began in earnest. Its light weight meant it was particularly well suited to serve aboard the giant dirigible airships, the Zeppelins, where the reserves of buoyancy were comparatively small; fitting six Parabellums instead of six Maxims saved the weight of a crewman. The water-cooling system was essential, as gas – highly combustible hydrogen – was often vented from tanks close to the gun positions on top of the gas envelope. The barrel of an air-cooled gun, red-hot after prolonged firing, would be a fatal liability.

7.9mm Gewehre 88/S and 88/05

When World War One began, Gew. 88/05 were being stored for the *Landwehr* units that had not received Gew. 98, and the surviving Gew. 88/S were being held for the *Landsturm*. About 500,000 *Gewehre* 88, of all types, were still serving the German armies at this time.

A trio of elderly Landsturm *men, pictured in occupied Belgium, carry rifles with ejection-port covers, suggesting that they are Gew. 88/05 or 88/14.*

A magazine cover was developed in December 1914 to prevent dust, mud and sand getting into the action through the opening in the bottom of the magazine well. A clip-ejector mechanism was added to throw empty clips up and out of the action; previously, they had simply fallen downwards out of the magazine when the bolt was opened to expel the last spent case. The alterations to the magazine system were undertaken exclusively by the government arsenal at Spandau in 1915.

7.9mm Scharfschützen-Gewehr 98

The unexpected rise of trench warfare during World War One renewed interest in sniping which had been dormant since the American Civil War. Several hundred impressed *Jagdgewehre* (hunting rifles), mostly Mausers, were fitted with commercial Gérard, Goerz and Zeiss telescope sights, and a few *Gewehre* 98 were fitted experimentally with 4× Goerz 'Certar Kurz' for trials on the Western Front.

The sporting rifles chambered the *Patrone* 88 instead of S-Munition and so, to prevent accidents, a tin plate label was added to the right side of the butt. This was stamped with a silhouette of a *Patrone* 88 and the warning 'NUR FÜR PATRONE 88' above 'KEINE S-MUNITION VERWENDEN' ('only for Patrone 88' above 'unsuitable for S-Munition').

Sporting rifles often had delicate 'set' triggers that were not suited to arduous service. As soon as trials with modified infantry rifles had been successfully concluded, therefore, the *Gewehr-Prüfungs-Kommission* ordered 15,000 *Zielfernrohr-Gewehre* 98 – subsequently known as *Scharfschützen-Gewehre*.

Sniper rifles were issued on the scale of one to each company in the Bavarian infantry regiments and *Jäger-Bataillone*, rising to three per company by August 1916. Prussian snipers usually operated independently. Most men ultimately received K-Munition, which gave better long-range accuracy than the standard 'S' pattern.

The *Scharfschützen-Gewehr* 98 was a specially selected and finished Mauser, capable of great accuracy. Its bolt-handle was bent down, and a Goerz or Zeiss 4× telescope sight, fitted in two ring mounts, was offset to the left to allow the magazine charger to be used. The range drums of the optical sights were graduated for 200m, 400m and 600m in Bavaria, but from 100m to 1,000m in Prussia, Saxony and Württemberg.

Captured Rifles

The capture of French and Russian rifles in 1914 freed Mausers and *Reichsgewehre* from Reserve, *Landwehr* and *Landsturm* units. By 6 November 1914, the German Navy alone had received 10,000 French Lebels and 7,000 assorted Russian Mosin-Nagants. A report from the Baltic Station command (Kiel) noted in mid-November 1914 that the French rifles had been issued to the Kiel dockyard division and I. *Flieger-Abteilung*; Russian guns had gone to the dockyard and munitions-depot guards, as well as the airship detachment.

However, strategically important tasks such as the guard mounted over the Holtenau locks and the Prinz Heinrich bridge across the North Sea Canal were still entrusted to men armed with *Gewehre* 98.

13mm Balloon Guns

The British used 12-bore shotguns against German observation balloons in the first few months of World War One, firing bullets loaded with phosphorus, but little is known about German anti-balloon armament or the employment of the Austro-Hungarian 'Alder' or *Ballon-Patrone*.

Obsolescent single-shot Mauser actions could undoubtedly have accepted a more effective cartridge than their small-calibre successors, but the calibre of the Gew. 71 'anti-balloon guns' is usually listed as 13mm. Assuming they had not been adapted to fire shotgun ammunition, there seems no good reason to change from 11mm. Unfortunately, no gun of this type has yet been traced for examination and so the questions remain unanswered.

7.9mm Gewehr 88/14

Approximating to the earlier 88/05, these were hastily altered from Gew. 88/S between December 1914 and the summer of 1915. Crudely shaped charger guides were welded to the front of the receiver-bridge; the left wall of the receiver was cut away to enable cartridges to be pressed into the magazine well; and a groove milled across the face of the chamber allowed *S-Patronen* to enter the magazine satisfactorily. A sheet-steel cover blocked the opening in the bottom of the magazine, and the spring-loaded cartridge retainer was angled forwards. The magazine well was shortened and narrowed for the S-*Patrone*, with sheet-steel inserts welded in place to compensate for the absence of the original clip. Conversion work was carried out very hurriedly and the standard of finish is notably inferior to that of peacetime 88/05 conversions. It is believed that the Spandau factory altered about 75,000 *Gewehre* 88/14, adding a 2mm 'n' on the barrel and chamber-top. Many of these guns ended their days in Bavaria, where nearly 35,000 were still on the inventory in October 1918.

10 1915

AUSTRIA–HUNGARY

8mm Repetiergewehr Modell 93

Romania had ordered about 200,000 1893-pattern Mannlichers from Österreichische Waffenfabriks-Gesellschaft in 1913, immediately after the end of the Second Balkan War. However, only about 75,000 had been delivered when World War One began. The Austro-Hungarian authorities ordered the Steyr factory to convert guns in the course of production to accept the standard 8 × 50mm rimmed cartridge instead of the 6.5 × 53mm pattern. Barrels were re-bored, and the magazine box was enlarged to accept the M93 clip with its wide-bodied cartridges. Rifling was identical to that of the standard Mannlicher pattern. The backsight was changed, and a longitudinal groove was cut in the top of the hand guard to allow the sight-notch to be used at the lowest adjustments. *Adaptierte Ausgebohrtes rümanisches Gewehre* (altered bored-out Romanian rifles) were marked 'STEYR 1914' on the left side of the receiver and had Austrian acceptance marks (for example, 'W-n [eagle] 15'). Some guns retained the Romanian crown over 'Md 1893'

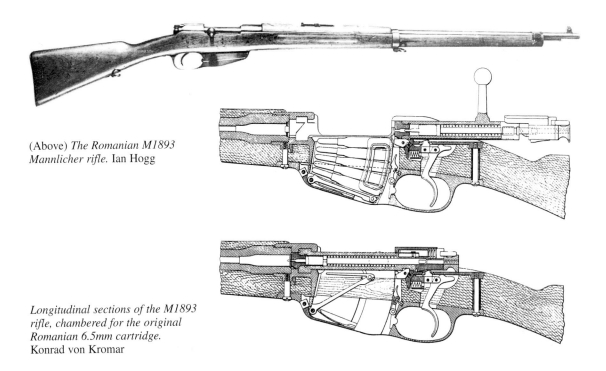

(Above) *The Romanian M1893 Mannlicher rifle.* Ian Hogg

Longitudinal sections of the M1893 rifle, chambered for the original Romanian 6.5mm cartridge. Konrad von Kromar

marks on top of the chamber, but all were accompanied by standard Romanian-type knife bayonets.

7.92mm Repetiergewehr Modell *14*

A decision to develop a new rifle to replace the straight-pull Mannlicher was taken at a meeting of the *Militär-Technische-Komitee* in mid-April 1915. Tried in the harsh conditions of the Eastern Front against the Russians, the 1895-pattern rifles had shown serious weaknesses. The barrel was deemed to be too light, the enveloping hand guard/fore-stock assembly was weak, and the lack of adequate primary extraction was worrisome if regular maintenance could not be guaranteed.

Although the replacement for the M95 is popularly believed to have been a Mauser, it was actually a 7.9 × 57mm Mannlicher-Schönauer with a rotary magazine and a charger-loading capability. However, the entry of Italy into World War One in May 1915 stopped work almost before it had begun.

7mm Repetiergewehr Modell *14*

About 5,000 Mauser rifles and associated bayonets ordered by Colombia were acquired from Österreichische Waffenfabriks-Gesellschaft in May 1915. Altered for the 1895-type sling, 3,000 were subsequently issued to volunteer riflemen in the Kärnten district. Identical mechanically to the Mexican pattern, the Mo. 1912 Colombian rifles were distinguished by a chamber-mark of a shield superimposed on four crossed national flags and surmounted by a condor carrying a scroll inscribed 'LIBERTAD Y ORDEN'. Charges on the shield consisted of pomegranate flanked by two cornucopias, above a Liberty Cap on a staff and two sailing ships separated by the Isthmus of Panama. The pre-1914 Colombian guns, unlike later deliveries, did not identify the country by name.

7.92mm Repetiergewehr Modell *13*

Shortages of weapons were partly overcome with the help of 66,000–72,000 (estimates vary) obsolescent *Gewehre* 88 supplied by Germany in 1915–16. Swivels were altered to suit Austrian practice, but the guns were otherwise used in their original condition.

Dating back to 1892 or earlier, they seem to have been clip-loaders (Gew. 88, Gew. 88/S) instead of the more effectual charger-loading adaptations (Gew. 88/05). Most of the guns were issued to the *k.k. Landwehr*, but so many problems were encountered in service that unwanted guns were sent to Turkey later in the war. The *Hirschfänger* M71 was regarded as the standard bayonet, but when supplies ran short, old *Pionierfaschinenmesser* were acquired instead. Typically Austrian bent-strip bayonets were also made in small numbers during the war.

A few *Repetierkarabiner* M13 (probably a mixed batch of Kar. 88 and Gew. 91) are said to have served with mountain units in the Tirol, but this claim has never been satisfactorily confirmed.

The M13 chambered the standard German 8 × 57mm cartridge, the cartons being marked (in black on yellow) '15 St. M. 13 scharfe Mauserpatrone 7.92mm'. Ammunition was usually made by Hirtenberg, but some consignments may have come from Germany.

8mm Maschinengewehr Modell *7/12*

Only twelve 8mm M7/12 and twenty-four 6.5 × 54mm M12 (ex-Greek) guns were serving the *k.u.k. Fliegerkompagnien* in July 1914; and only six of these thirty-six guns had been mounted in *Luftfahrtruppen* aircraft by February 1915.

7.92mm Madsen-Muskete

The declaration of war by Italy, on 23 May 1915, caught Austria-Hungary unprepared for alpine fighting. Realizing that light automatic weapons were in short supply, the war ministry purchased 632 Madsens from Denmark. Chambered for the 6.5 × 55mm round, these were re-bored in the Vienna arsenal for the German 7.9 × 57mm cartridge (known as '7.92mm' in Austria-Hungary), but the alterations took so long – teething troubles must have been appreciable – that issues were slow. Some guns went to Army units fighting in the south; about fifty were issued to the *k.u.k. Fliegerkompagnien*, but the weapons, unsuccessful in air service, were withdrawn in February

1917. It is assumed that the ground-based guns survived until the end of World War One.

Impressed Sporting Guns

Many sporting rifles were pressed into service in 1915 for sniping, training, patrol duties and similar unimportant roles, thus freeing service weapons for front-line use.

The original *Reichsgewehr*-type Mannlicher rifle had not been popular commercially, as few European sportsmen liked its clip-loaded magazine, but the advent in 1904 of the Mannlicher-Schönauer was greeted with greater enthusiasm. It was well made, sturdy, easy to use and comparatively inexpensive.

The earliest examples were identical mechanically to the Greek M1903 service rifle, though greater care was taken over surface finish. Chamberings were initially restricted to the 6.5 × 54mm rimless cartridge.

Most of the guns were made as *Repetier-Pirschstutzen* (short rifles), with full-length stocks, but long-barrel half-stock *Repetier-Pirschbüchsen* were also available. The military-style action could be loaded from a charger, a safety-catch lay on top of the cocking-piece, and the bolt-handle was of flattened spatulate form inspired by the German Kar. 88.

The butts usually had a straight comb, a small oval cheek-piece and a rounded pistol-grip. The grip and fore-end were chequered, a steel nose-cap was fitted, and a sling ring was held by a small collar on the barrel and a bolt through the fore-end. Traps in the butt-plate could hold a cleaning rod and additional cartridges.

The most popular form of backsight, the *Klappvisier*, had a standing block and two small folding leaves. Guns were often fitted with double 'set' trigger units, but single-trigger patterns were also made. A selection of different chamberings appeared in 1905–12, but most of the guns impressed for military service handled the German 8 × 57mm sporting cartridge.

GERMANY

9mm Pistole 08

A mystery is presented by the absence of authenticated Erfurt-made Parabellums dating from 1915. Writing in *Die Pistole 08*, Joachim Görtz has suggested that, when World War One began, supplies of long-barrel Parabellums were only just beginning to reach the field artillery. If the Erfurt factory had been ordered to complete all 1914-dated parts in this form, this would account not only for

Men of the cruiser Königsberg, *marooned in the Rufiji Delta, prepare to leave to join the German forces fighting in southern Africa, 1915. The Mauser rifles and Parabellum pistols were taken from the shipboard armoury.*

the '1914' date invariably encountered on the Erfurt-made LP08, but also for the complete absence of 1915-date standard *Pistolen* 08.

9mm Mauser-Selbstladepistole *C/96*

A contract – possibly open-ended – was let with Waffenfabrik Mauser in 1915 for at least 150,000 C/96 pistols adapted to chamber the standard 9mm service-pistol cartridge. This order was still incomplete by the time of the Armistice in November 1918, the highest known number being 141,007. The guns conformed to the standard 'NS' pattern, but had backsights graduated from 50m to 500m and grips carved with a large '9' to show that they were chambered for the 9mm *Pistolenpatrone* 08 instead of the 7.63mm Mauser cartridge (*see* Part Three: 'Germany: handguns').

Handgun Shortages

Rifles and machine-guns were accorded priority, and so the general shortages of weapons soon extended to handguns. Pistols had never been required in large numbers, and the manufacture of service-pattern guns was largely confined to

Deutsche Waffen- und Munitionsfabriken, Mauser and the government factory in Erfurt.

Although Deutsche Waffen- und Munitionsfabriken and Erfurt made 1.63 million Parabellums between them prior to November 1918, both agencies had more pressing business in 1915, forcing the *Kriegsministerium* to recruit new agencies to ensure supplies of handguns. Alternative sources of *Pistolen* 08 were investigated – the Bavarian state rifle factory in Amberg, Waffenfabrik Mauser, and even Bosch in Stuttgart – but only Anciens Établissements Pieper of Herstal-lèz-Liége was recruited (the Pieper family had ties in Germany), supplying hold-opens, strikers, magazine followers, safety catches, triggers, trigger-plates and other small parts directly to the Erfurt ordnance factory.

A surprising variety of handguns eventually found its way into official service. Few of these *Behelfspistolen* (substitute patterns) were particularly powerful, but they did free Parabellums for front-line service. Material published later in the war (*see* Part Three: '1917') indicates acquisition of a broad range of commercial designs, including tiny blowbacks seized after the invasion of Belgium.

(Left) *The 9mm Mauser C96 pistol was accepted for service during World War One. Most guns had a large '9' impressed in the grips.* Thompson D. Knox

(Right) *The Mauser was customarily accompanied by a special shoulder stock-holster unit. This 7.63mm gun was captured from a German artillery officer in August 1918.*

Officers had always purchased small-calibre weapons for personal defence, but these did not normally bear ordnance markings of any type.

Among the guns that deserve more than a passing mention is the 7.65mm Dreyse, patented in 1906–07 by Louis Schmeisser and made by Rheinische Metallwaaren- und Maschinenfabrik of Sömmerda. Introduced *c.*1908, most (if not all) of the initial 1,000-gun production run went to the gendarmerie in Saxony. The guns customarily display 'K.S. Gend.' and an issue number corresponding with the serial number of the gun. The Dreyse had sold in substantial quantities prior to 1914 – the Berlin police used guns marked 'K.P.P.B.' – and continued to be made in quantity throughout World War One.

The basic gun consisted of a fixed frame with a separate slide/breech-block assembly which recoiled backward on firing. The design of the retraction grips at the front of the slide gives a clue to the gun's age; the original version had vertical grooves confined to the slide, but these were subsequently extended downwards over the frame, became diagonal on the slide alone, and finally

took diagonal form on a squared backing. Dreyse pistols are typically 160mm long, have 93mm barrels and weigh about 710g empty; their magazines hold seven rounds.

The maker's mark originally read simply 'Rheinische Metallwaaren- & Maschinenfabrik' over 'ABT. SÖMMERDA' on the left side of the frame. This was joined by 'DREYSE • 1907' (subsequently simply 'DREYSE'), and eventually changed to 'DREYSE' above 'RHEINMETALL ABT. SÖMMERDA'. Military inspectors' marks have been found on guns numbered as high as 195,000.

The Dreyse was made in small numbers in 6.35mm and 9mm, the former being a pocket pistol whereas the latter represented an unsuccessful attempt, dating from 1912–13, to enlarge the basic design for the standard 1908-type service cartridge.

Designed by Heinz Zehner and patented in Germany in 1912, the 7.65mm seven-shot Sauer was another pre-World War One design that had seen police use. Introduced commercially in 1913, it was a blowback design with the recoil spring

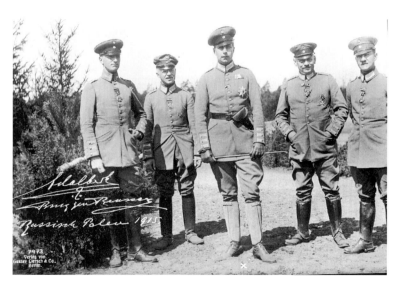

In addition to the regulation pistols, all types of substitutes were impressed into military service. The holster worn by Prince Adalbert of Prussia clearly contains a very small pistol, perhaps a 7.65mm Sauer or something similar, while the mountaineering officer on the left in the second picture carries something a little more substantial. The magazine pouch on the holster-face and the cleaning rod on the spine indicate that it is not a regulation Parabellum.

The 7.65mm Langenhan 'FL'-Selbstlader was another of the Behelfspistolen *accepted for service in 1915–18.*

encircling the barrel. The striker-fired Sauer had a good reputation for accuracy and was undoubtedly better made than many of its rivals. Several minor variants were made, differing in safety arrangements and dismantling systems; the first guns had a separate magazine-safety mechanism, but this was rapidly abandoned. By 1918, at least 85,000 7.65mm Sauers had been made – though not all had been purchased officially. The guns were 144mm long, had 75mm barrels and weighed 575g empty.

The 7.65mm 'FL-*Selbstlader*', promoted by Friedrich Langenhan of Suhl, was a wartime introduction distinguished by a separate breech-block held in the slide by a pivoting yoke and a large knurled-head nut. The right side of the slide was weakly constructed, owing to the size and position of the ejection port, and the lock-nut showed a distressing tendency to work loose after only a few rounds had been fired – with potentially disastrous consequences if the yoke disengaged and the breech-block flew backward out of the gun.

Most 7.65mm FL pistols display military inspectors' marks, and it is suspected that virtually all of them went to the German Army; production reached 85,000. Changes made to the ejection port and the trigger/disconnector mechanism were acknowledged by the incorporation of a second

DRGM number in the slide mark after about 13,000 guns had been made. The pistols were about 168mm long, had 105mm barrels and weighed 670g. Magazines held eight rounds.

The 7.65mm *Modell* 4 made by Waffenfabrik Carl Walther of Zella-Mehlis was an enlargement of the company's blowback pocket pistols. Amongst the most popular substitute designs, about 75,000 were supplied to the German armed forces in 1915–16 and many others were sold privately during World War One. The most obvious feature was a light sheet-steel extension attached inside the front of the slide with a bayonet joint, though the ejection port was on the left side of the slide. A typical *Modell* 4 measured 151mm overall and weighed about 550g with an empty seven-round magazine; the barrel was 85mm long.

The Walther *Modell* 6 was an enlarged *Modell* 4 adapted to chamber the 9mm Parabellum cartridge, but the absence of a mechanical breech-lock was viewed with suspicion and only about 1,000 were made.

The 7.65mm Beholla, made by Becker & Hollander (45,000) and August Menz of Suhl (15,000), seems to have owed its origin – as did the 7.65mm 'FL-*Selbstlader*' – to the *Hindenburg Programm* of 1915. Simple, robust and serviceable, the blowback Beholla was subsequently made by several contractors after World War One had finished. Its only major flaw was a quirky dismantling system. The 7.65mm Beholla pistol was 140mm long, had a 76mm barrel and weighed 640g. The magazine held seven rounds.

The 7.65mm Jäger-Pistole, the work of Franz Jäger of Suhl, was a particularly interesting gun. The gun was a simple blowback, with a frame that was assembled from two sturdy pressings held apart by pinned straps acting as spacers. The machined breech-block was held in a pressed-steel slide, and the return spring was concentric with the barrel. It has been estimated that about 12,000 Jäger pistols were made, including some for commercial sale. Most surviving examples are marked 'JÄGER-PISTOLE. D.R.P. ANGEM.' in a single line on the left side of the frame.

The 7.65mm blowback Mauser pistol, well made but comparatively expensive, was also approved for military service. The first guns had a distinctive hump-backed slide, but this had been replaced by a straight-top design before World War One began. Analysis of serial numbers suggests that the Mausers were offered commercially until 1915, but that virtually all production thereafter went to the German Army; guns numbered from about 73,000 up to 180,000 are customarily found with inspectors' marks. The 7.65mm Mauser is 153mm long and weighs about 600g; the barrel measures 87mm and the magazine holds eight rounds.

Details of *Behelfspistolen* will be found in many specialist books – in particular, in the painstaking work of Jan Still (*see* Bibliography).

Machine-Gun Issues

One of the most important events in 1915 was the wholesale expansion of the German Army's machine-gun units. *Feldmaschinengewehrzüge* or *Maschinengewehr-Ergänzungszüge* (machine-gun sections) of three or four guns apiece were formed in quantity; some regiments raised additional machine-gun companies; and the first *Maschinengewehr-Scharfschützen-Trupps* (machine-gun marksman sections) were created from picked men.

By early April 1915, Deutsche Waffen- und Munitionsfabriken was said to be making 1,400 Mauser rifles, 700 Parabellum pistols, ten machine-guns and two million small arms cartridges daily, together with 10,000 grenades, 5,000 fuzes and a large number of shell-cases. The company's total

The 7.65mm Mauser blowback pistols were accompanied prior to 1914 by an essentially similar, but substantially more powerful delayed-blowback design of the type pictured here. A few of these guns, purchased by individual officers, saw limited service during World War One. Henk Visser

machine-gun production for the 1915–16 financial year amounted to about 3,950.

7.9mm *Parabellum*-Maschinengewehr

Fitted with a large-diameter slotted barrel casing, an air-cooled LMG14 was the first synchronized machine-gun to be fitted to a German aeroplane. A single gun was fitted experimentally to a Fokker M.5K/MG monoplane in May 1915, though testing may have been confined to ground-firing. Parabellum machine-guns firing through the propeller arc were subsequently fitted to the Fokker 6.1 *Eindecker* ('Monoplane') until the perfected LMG08 became available (*see* pages 188–9).

The Parabellum-type LMG14, with slotted barrel casing. Ian Hogg

Zu den Kämpfen in Russland.
Maschinengewehre im Schützengraben
vor Warschau in Tätigkeit.

(Above) *An MG08 crew – possibly from the machine-gun company of 26. Infanterie-Regiment – pose in a ruined village in northern France in 1915, clearly not under fire!*

(Left) *An MG08 crew demonstrate their weapon in a trench 'before Warsaw', though the absence of protection suggests that the front-line is some way distant.*

7.9mm Maschinengewehr *08*

The Maxim machine-gun had a steady and efficient action, but its cyclic rate was not much more than 300rd/min. During World War One, therefore, a *Rückstossverstärker* S (recoil booster) was added to increase the fire-rate to about 450rd/min by diverting propellant gas emerging at the muzzle to increase the rearward thrust on the barrel.

Production of German Maxims was initially entrusted to Deutsche Waffen- und Munitionsfabriken in Berlin and the government rifle factory in Spandau, but new contractors were recruited from 1915 onwards to make the MG08/15 (qv).

Rifle Issues

The long Mauser rifle was being carried by the regular infantry regiments and the greater part of the Reserve in August 1914, but supplies were not large enough to equip many of the units raised on mobilization even though Deutsche Waffen- und Munitionsfabriken was making 1,400 rifles and two million rounds of 7.9mm ammunition daily by 1915.

According to the regimental history, the men of Magdeburgisches Infanterie-Regiment Nr 66 carried *Gewehre* 98 until 1916, when they were replaced first by ex-Russian Mosin-Nagants and secondly by *Gewehre* 88/05 before another consignment of *Gewehre* 98 appeared.

Shortages were made good partly by accelerating production of the *Gewehr* 98, partly by issuing *Gewehre* 88/S, 88/05 and 88/14 from store, and partly by impressing captured rifles.

Published in 1915, *Kurze Beschreibung der an Ersatztruppen und Rekrutendepots verausgabten fremländischen Gewehre* ('A short description of the foreign rifles given to supplementary units and recruiting depots') lists a surprising variety of equipment. This included the British Mk I and Mk III SMLE; the Canadian Ross 'M1910'; US single-shot Remingtons taken from the French; and a Peabody rifle of unknown provenance.

The Belgians contributed the Albini-Braendlin, the Comblain and 1889-type Mausers; the French gave the Mle 66 Chassepot, Mle 74 Gras, Mle 78 Navy Kropatschek, Mle 86/93 Lebel, Mle 90 Berthier carbine and Mle 92 Berthier musketoon. Italy provided the Vetterli-Vitali and Mannlicher-Carcano rifles. Russia contributed the Berdan and a selection of Mosin-Nagants. Austria-Hungary gave the M95 Mannlicher rifle and *Stutzen*. The Dutch Beaumont rifle, a Dutch Remington gendarmerie carbine and – perhaps surprisingly – Dutch 1895-type Mannlichers were also listed.

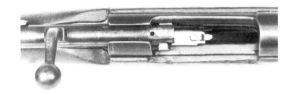

(Above) The Italian Mo. 1891 rifle, with the bolt open. Ian Hogg

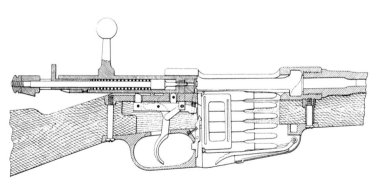

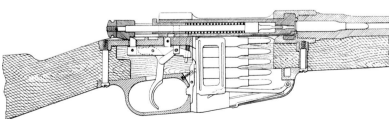

Longitudinal sections of the Italian Mo. 1891 Mannlicher-Carcano rifle. Konrad von Kromar

Many clearly came from captures, including most of the Belgian, British, Canadian, French and Russian guns. The source of the older Dutch guns was presumably the stocks of major German arms dealers such as A.L. Frank and Benny Spiro. The Dutch Mannlichers are the most difficult to assess; were they purchased from the Hembrug factory when World War One began, even though the Dutch professed to be neutral?

The wholesalers had colossal stocks of military-surplus weaponry; in 1911, A.L. Frank alone had 250,000 Austro-Hungarian Werndl rifles and 42,000 Italian Vetterli rifles and musketoons. There were also enough Belgian Albini-Braendlin and Dutch Beaumont guns in Frank's warehouses to equip entire regiments. What happened to some of these is conjectural, although the absence of references to Werndls in German literature suggests

that they were returned to Austria-Hungary. This made good sense; the Germans did not have appropriate cartridge-making facilities, whereas the Austro-Hungarians still had 11mm ammunition in store. The same fate may also have befallen the thousands of Wänzl rifles in Frank's possession.

Evidence provided by the vast numbers of post-cards that survive from the pre-1918 era is invaluable, but photographic studios (particularly in garrison towns) may have kept damaged or obsolete weapons as props. However, though this may have contributed some of the most bizarre combinations of guns and soldiery, it does not explain the extraordinary range of modifications made to bayonets.

Bayonet production was clearly unable to keep pace with the output of rifles, creating a need that had already been answered by ordering one million extra *Seitengewehre* 14. Manufacture of the so-called *Ersatz* patterns (*Aushilfs-Seitengewehre 88/98*) and adaptation of old weapons began early in 1915. The ingenuity of these conversions is best appreciated by studying Anthony Carter's monumental four-part work, *German Bayonets* (*see* Bibliography).

Ransacking dealers' stocks often obtained large numbers of bayonets without appropriate rifles; A.L. Frank, for example, had 1,800 Chassepot rifles in 1911 – but nearly 80,000 Mle 66 bayonets! Interestingly, 1,400 of the bayonets had already been altered 'for the Gewehr 71'. The relevant totals for the Gras were 6,000 rifles and 39,200 bayonets. As 3,200 Gras bayonets had also been adapted to fit the old German Mausers, it is clear that work now customarily credited to the exigencies of war had sometimes been undertaken several years previously.

Other sources of rifles were overruns on large contracts, or cancellations. About 10,000 Mauser rifles stored in Oberndorf since the Chinese revolution of 1911 were reissued in 1915 to reserve infantry regiments 261 and 262; some may also have been used by a *Jäger* unit. Practically identical to the Gew. 98, the Chinese guns had short 'export' nose-caps and accepted bayonets with conventional muzzle rings.

A few 1907-pattern Haenel rifles were also issued to the German Army, though the total may

A page from the 1911 ALFA catalogue, showing some of the Mannlichers that were available from stock.

not have exceeded 1,000. Based on the *Reichsgewehr* (Gew. 88), these guns had pistol-grip stocks, bolt-handles turned downwards, and a patented quick-release magazine floor plate. Originally offered to China in 6.8 × 57mm, World War One issues were apparently re-bored to handle 7.9mm S-Munition – even though they were still marked '6,8mm Mod. 07' on the left side of the receiver. A boss on the nose-cap accepted the 'muzzle ring' of the bayonet, and the attachment lug lay on a plate extending backwards beneath the stock.

Units of the Württemberg Army serving in the vicinity of the Mauser factory in Oberndorf used a few thousand 9.5mm Turkish M1887 rifles. Left part-complete in the factory when the 1890-pattern rifle had been substituted, these had been completed in the 1890s and kept in store.

The Germans also made limited use of Austro-Hungarian Mannlichers, though the circumstances are still unknown. These guns may have been issued to German units serving alongside Austro-Hungarians on the Southern Front. But it is also possible that some were acquired by the Bavarians, probably in 1916, when the Amberg factory was unable to meet demands for *Gewehre* 98. Though many of the German-made 1888- and 1895-type knife bayonets have Prussian-style 'crown/W' ciphers, the Bavarians ordered M12 Steyr-Hahn pistols in 1916 when the Prussians were unable to

supply sufficient Parabellums and a similar course may have been taken with rifles.

Problems in the German Army were mirrored in the Navy, where it was soon clear that the existing inventory of Gew. 98 could not arm the ever-increasing numbers of men mobilizing for action. From May 1915 until the end of World War One, for instance, only 16,959 Gew. 98 and 1,700 Kar. 98 AZ were delivered to Kiel dockyard for the entire *Ostsee* (Baltic) district.

The shortfall was answered by issuing *Beutegewehre* – captured rifles – generally Russian Mosin-Nagants and French Lebels. On 14 April 1915, Maxim machine-guns had been withdrawn from U-boats (which had been using them to destroy floating mines) in return for two Russian rifles. An inventory taken on the Baltic Naval Station (Kiel) in August included 8,726 Mosin-Nagants: 2,648 serving aboard warships and auxiliaries, and a further 6,078 on land service. By the end of 1915, virtually every small warship had three guns for anti-mine duties. Even as late as February 1918, Kiel still had more than 5,000 *Beutegewehre*, with Mosin-Nagants predominating.

7.62mm Aptierte Russisches Gewehr *M91*
Substantial quantities of Russian Mosin-Nagants were altered in German service. Some apparently had their magazines altered and the barrels bored-

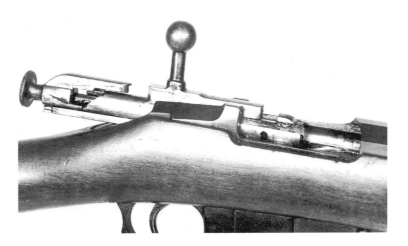

The open action of the Russian 1891-type Mosin-Nagant rifle.
Ian Hogg

A longitudinal drawing of the action of the Mosin-Nagant.

(Left) *The Mosin-Nagant was issued to the men of many second-line units, including Driver Gottlob Hehr attached to a field hospital of 5. Armee-Korps. The postcard is postmarked February 1916.*

(Above) *The German Navy also used Mosin-Nagants in large numbers. This group from* I. Matrosen-*Division, pictured in Kiel in 1918, carry Russian rifles with muzzle adaptors and a selection of substitute or* Ersatz *bayonets.* H. Gregory Engleman

out to fire the 7.9×57mm cartridge, but most seem to have been used with captured Russian ammunition. However, changes were made to enable a shortage of bayonets to be overcome.

Although at least one special socket bayonet was developed, a more popular method was to cut back the fore-end and attach a sleeve-like adaptor to the muzzle to accept a standard *Aushilfsseitengewehr* 88/98. A similar system was tried with captured French Mle 86/93 Lebels, but Mosin-Nagant examples now predominate and it is suspected that production for French guns was very small. Moritz Magnus the Younger of Hamburg apparently designed the adaptors in the spring of 1915.

Altered Mosin-Nagants customarily had an eagle within a 'DEUTSCHES REICH' cartouche on the side of the butt, struck over the original Russian marks.

Machine-Guns

7.9mm Madsen-Muskete

Danish-made Madsen light machine-guns were acquired in small numbers in the spring of 1915. These recoil-operated air-cooled weapons, with top-feed box magazines holding twenty-five rounds, first saw service in the Champagne district in September – with three specially-raised *Musketen-Bataillone* – but they were not particularly successful and were withdrawn after the Somme battles of the second half of 1916. A *Musketen-Bataillon* consisted of three companies, each armed with thirty light machine-guns operated by four-man squads.

A few Madsens were taken from the Russians, chambered for the rimmed 7.62 × 54mm cartridge. These are said to have been confined largely to air service, owing to the problems of supplying ammunition, but occasionally served as ground guns.

It is unclear whether the failure of the Madsens was due to problems arising from the 7.9 × 57mm adaptation or simply because they did not fit German tactical doctrines. However, the guns could not sustain fire long enough to replace belt-feed Maxims, and were subsequently reissued to *Gebirgsjäger-Maschinengewehr-Abteilungen* (mountain troops). These units initially served in the Vosges, but were dispatched to the Carpathians and the Balkans in 1916 to bolster Austro-Hungarian resistance. The light weight and box-magazine feed of the Madsen were particularly advantageous in highland areas.

7.9mm Dreyse-Maschinengewehr Modell 1915 (MG15)

This machine-gun had its origins in a patent granted to Louis Schmeisser in 1909. Schmeisser had previously designed the Bergmann machine-gun (qv), but the original patents were still valid and something different was needed by his new employers, Rheinische Metallwaaren- und Maschinenfabrik.

The MG15 may also owe its development to the competition undertaken in 1908 to find a perfected machine-gun for the German Army, though the 1908-pattern 'light' Maxim was successful.

7.9mm *Dreyse-Maschinengewehr Modell 1915* (MG15)

Synonym:	German Dreyse machine-gun
Adoption date:	1915
Length of gun:	about 1,085mm (42.7in)
Weight of gun:	14.5kg (32lb) empty
Weight of coolant:	4kg (8.8lb)
Weight of mounting:	31.3kg (69lb)
Barrel length:	704mm (27.72in)
Chambering:	7.9 × 57mm, rimless
Rifling type:	four-groove, concentric
Depth of grooves:	0.15mm (0.006in)
Width of grooves:	4.5mm (0.173in)
Pitch of rifling:	one turn in 240mm (9.45in), RH
Feed type:	100- or 250-rd fabric belt
Weight of loaded belt (250-rd type):	7.25kg
Front sight:	open barleycorn
Backsight:	tangent-leaf type
Backsight setting:	*minimum* 400m *maximum* 2,000m
Rate of fire:	550 rd/min (without booster)
Muzzle velocity:	860m/sec (2,820ft/sec)
Bullet weight:	9.85g (152gn), 'S' type

Dreyse machine-guns were offered commercially from 1912 onwards, but lacked a designation and were usually marked simply 'DREYSE'; terms such as 'Model 1912' or 'Model 1913' are apparently inappropriate.

The earliest water-cooled machine-guns were mounted on light tube-leg tripods fitted with small pressed-steel wheels. Modified patterns were submitted to the authorities when World War One began, and Rheinische Metallwaaren- und Maschinenfabrik are said to have been given a contract to supply guns in quantity in 1915. Modified by the addition of a bracket for the 08/15 bipod, and with the option of a shoulder stock, the water-cooled MG15 seems to have been issued in small numbers as a rudimentary light machine-gun in 1916.

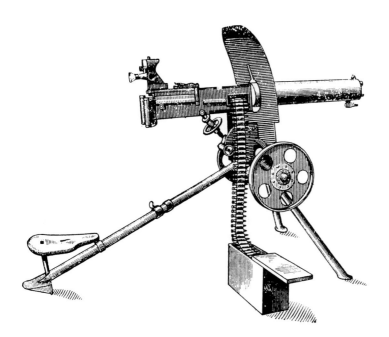

*A water-cooled Dreyse
machine-gun on its tripod mount.*

However, as the company was soon switched to making MG08/15 Maxims, it can probably be inferred that the Dreyse did not work well enough under battle conditions. Although several thousand Schmeisser-designed machine-guns had been made, they were subsequently sent to Turkey for service in Mesopotamia.

The Dreyse was easy to load or to clear once jammed, but had limited reserves of power. A bolt-driven pawl on top of the feed way moved the belt one cartridge leftward as the action closed; spring-loaded claws withdrew a cartridge from the fabric belt as the bolt ran back, then lowered the cartridge to be pushed forward into the chamber on the return stroke. However, although a pivoting lever was employed to accelerate the bolt as it moved backwards, the Dreyse did not perform well under adverse conditions.

The compact rotating-hammer firing system was efficient, but a flimsy strut, pivoted in the bottom of the receiver behind the bolt, sufficed to lock the bolt and barrel extension until disengaged by cams during the recoil stroke. This arrangement had an affinity with the so-called 'M1903' Mannlicher pistol,

which was also generally reckoned to be a weak design. It had none of the elegant simplicity of the Schmeisser-designed Bergmann machine-gun.

7.9mm Mauser Selbstlade-Gewehr Modell 15

This was a development of the C06/08 rifle, patented in October 1906 and originally given field trials in 1907. When the gun fired, the barrel and receiver slid backwards while locking bars were cammed into the sides of the receiver to release the breech-block.

The action was cumbersome and unreliable, and the slender locking bars were particularly prone to breakage, but Mauser designers were doggedly persisting with the flap-lock when World War One began. Consequently, a few half- and full-stocked guns were issued in 1915.

The full-stocked rifles had Gew. 98-pattern nose-caps, accepted standard bayonets, and were tested in the trenches of the Western Front. The half-stock patterns were initially used for air combat, but experience showed the Mondragon (qv) to be preferable and the Mausers were rapidly withdrawn. Surviving guns seem to have been reissued to the German Navy.

7.9mm *Mauser Selbstlade-Gewehr Modell* 15	
Synonyms:	*Flieger-Gewehr*, Mauser automatic rifle M1915
Adoption date:	1915
Length of gun:	1,245mm (49.02in)
Weight of gun:	4.725kg (10.42lb) empty
Barrel length:	675mm (26.57in)
Chambering:	7.9 × 57mm, rimless
Rifling type:	four-groove, concentric
Depth of grooves:	0.15mm (0.006in)
Width of grooves:	4.5mm (0.173in)
Pitch of rifling:	one turn in 240mm (9.45in), RH
Magazine:	detachable box
Feed:	chargers or single rounds
Magazine capacity:	10 or 20 rounds
Front sight:	open barleycorn
Backsight:	tangent-leaf
Backsight setting:	*minimum* 400m *maximum* 2,000m
Muzzle velocity:	830m/sec (2,725ft/sec)
Bullet weight:	9.85g (152gn), 'S' type

7.9mm Leichtes Maschinengewehr Modell 1915 (LMG15)

The Bergmann of World War One had a lengthy pedigree, but no prior success. Its basic design had been patented in the name of Theodor Bergmann in 1901, even though Louis Schmeisser had been responsible for its design.

The perfected or 1910-pattern Bergmann machine-gun was a greatly refined version of the guns tested earlier in the century. The locking system was retained, and the gun still fired from the closed-bolt position. However, the feed was altered to a more robust design driven by the recoil of the barrel and barrel extension. This proved to be capable of lifting a greater weight of ammunition than the original bolt-driven feed.

Changing the feed from the original 'push-through' belts to the standard 'withdrawal' Maxim pattern, though retrograde in many ways, allowed the Bergmann and the Maxim to share the same ammunition. The Austrian Keller-Ruszitska disintegrating-link metallic belts could also be used if appropriate.

The breech of the Mauser Selbstlade-Karabiner, *with the trigger-guard unlatched.* Joseph J. Schroeder

(Left) *The case for the Mauser Selbstlade-Karabiner and its magazines.* Joseph J. Schroeder

The Bergmann had a fire-rate of 480–600rd/min, which was considerably greater than the 300rd/min of the standard Maxim. This was largely due to the efficiency of the locking mechanism, which had a comparatively short travel.

Water-cooled Bergmanns were used in small numbers during World War One, mounted on *Schlitten* 08 with suitable adaptors. An additional mounting point beneath the receiver accepted the elevating mechanism, though the gun mount was a clamp-and-gimbal around the water jacket.

The history of the LMG15 remains obscure. A suggestion has been made that, in 1915, contracts had been given to Rheinische Metallwaaren- und Maschinenfabrik and Bergmanns Industriewerke to supply the German Army with heavy and light machine-guns respectively. The Dreyse machine-gun (qv) was unsuccessful, and Rheinische Metallwaaren- und Maschinenfabrik was eventually ordered to make Maxims; Bergmann had better fortune.

The LMG15 was little more than an M1910 without the water jacket. The barrel was protected by a small-diameter slotted sheet-steel casing, the spade grips were replaced by a pistol-grip and trigger-guard beneath the receiver under the feed way,

---

7.9mm *Leichtes Maschinengewehr* Modell 1915 (LMG15)

Synonyms:	German Bergmann light machine-gun, M1915
Adoption date:	1915
Length of gun:	1,121mm (44.13in)
Weight of gun:	12.875kg (28.38lb) empty
Weight of mounting:	3.5kg (7.72lb)
Barrel length:	716mm (28.19in)
Chambering:	7.9 × 57mm, rimless
Rifling type:	four-groove, concentric
Depth of grooves:	0.15mm (0.006in)
Width of grooves:	4.5mm (0.173in)
Pitch of rifling:	one turn in 240mm (9.45in), RH
Feed type:	200-rd non-disintegrating metal-link belt
Weight of loaded belt:	5.5kg (12.13lb)
Front sight:	open barleycorn
Backsight:	a pivoting bar and slide
Backsight setting:	*minimum* 400m *maximum* 2,000m
Rate of fire:	550 rd/min
Muzzle velocity:	860m/sec (2,820ft/sec)
Bullet weight:	9.85g (152gn), 'S' type

The Bergmann LMG15 n/A on its miniature tripod. Ian Hogg

and a small shoulder plate – hardly qualifying for the description 'butt' – was attached to the back of the receiver. A folding-leaf sight lay on the receiver and a bracket for a belt-box could be attached to the right side of the gun.

Whether the Bergmann began its career in the air or on the ground is still open to question. In the summer of 1916, however, more than 100 *leichtes Maschinengewehr-Abteilungen* (light machine-gun sections) were formed in the Döberitz machine-gun school. These are known to have been issued with 'light Bergmann machine guns' that could only fire 300 rapid-fire rounds before the barrel overheated. They were clearly air-cooled, as the water-cooled Bergmann had the same ability to sustain expend as the 1908-type Maxim. As development of a water-cooled light machine-gun (the MG08/15) was already complete, the evidence – albeit circumstantial – suggests that a use was being found for displaced air service guns.

Most of the *leichtes Maschinengewehr-Abteilungen* were sent to the Eastern Front, where air-cooling presented fewer problems than water-cooled systems in sub-zero conditions. The Bergmanns also had barrels that were easy to change, but they were withdrawn from front-line service when the MG08/15 became available in quantity in 1917.

The original Bergmann light machine-gun fired from an open bolt, but accuracy was poor in the ground role and a modified pattern – the LMG15 *neuer Art* (nA: new pattern) – was substituted.

The action was efficient enough, but gave problems in aerial combat, when the *g*-forces of turns, spins and dives played havoc with the feed mechanism. The root of the problem lay in the belt system which, when the mechanism ran back, moved a slide mechanically to the right to engage the next round. This movement compressed a spring. When the bolt and barrel extension moved back to battery, the slide was moved leftward partly mechanically and partly by the pressure exerted by the 'loaded' spring. Under certain conditions, particularly violent leftward manoeuvres, the spring did not assert its full power and a jam could occur.

Bergmann light machine-guns shared the belt-drum developed for the MG08/15, which, owing to its simplicity, could be made in quantity long before the first of the guns appeared.

7.9mm Maschinengewehr Modell *1908/15* (MG08/15)

Although the Maxim-pattern MG08 was sturdy and reliable, it was heavy and difficult to move. The gun alone weighed about 26.5kg, and the perfected sledge mount, the *Schlitten* 08, contributed

The Maxim-type MG08/15. Ian Hogg

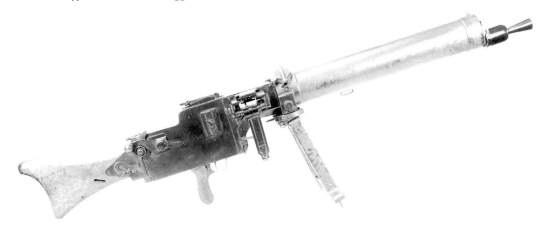

an additional 32.5kg. Stability conferred by weight suited the MG08 to static positions, but was of little help when close-range supporting fire was needed in attacks across the mud, blood and thunder of 'No Man's Land'.

When the first Lewis light machine-guns were captured on the Western Front and proved their value in trials, the *Gewehr-Prüfungs-Kommission* was instructed to find a lightweight weapon for the German forces. Dreyse, Bergmann and Parabellum guns were tested, but none proved acceptable.

The Danish Madsen proved to be reliable and, as it fed from a detachable box magazine on top of the receiver, was also exceptionally portable. However, though sizeable numbers were purchased from the Dansk Rekylriffel Syndikat of Copenhagen, for

use in terrain where belt-feed Maxims were unsuitable, buying a Danish-made gun was not the answer. Dansk Rekylriffel Syndikat was following an actively neutral course – selling guns to Allies and Central Powers alike – but Denmark was vulnerable to Allied pressure if too much war *matériel* was diverted to Germany.

Adopting a Bergmann or a Dreyse, unproven on the battlefield, would fail to make the best use of existing facilities by retaining as many Maxim components as possible. The outcome of the trials was the MG08/15, a lightened refinement of the basic MG08. Credited to a team of Army technicians led by *Oberst* von Merkatz of the *Gewehr-Prüfungs-Kommission*, the MG08/15, which reached the front-line in quantity at the end of

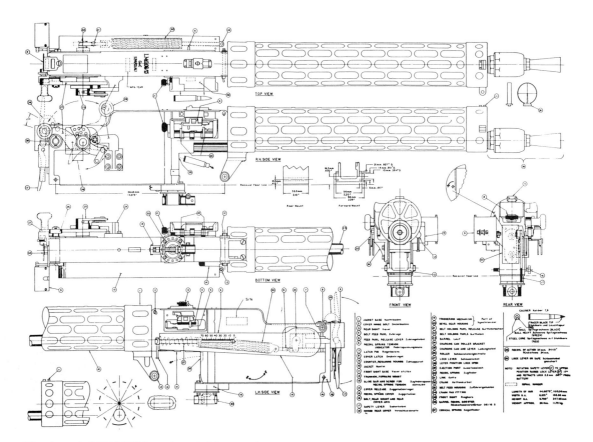

The air-cooled LMG08/15 was intended for aerial use.

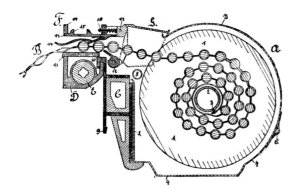

The Kasten *16, or cartridge-belt drum for the* MG08/15.

1916, was little more than a lightened MG08 with a wooden shoulder stock, and a pistol-grip and trigger mechanism beneath the receiver.

The diameter of the water jacket was reduced from 10cm to 9cm, reducing coolant capacity from 4lt to 3lt, and the thickness of the side walls of the receiver was reduced by about a quarter to save weight. The ejector tube was eliminated, the receiver was cut-down behind the backsight base, and a simple bipod replaced the 1908-type sled. The safety-catch on the upper left side of the pistol-grip unit was operated manually, unlike the automatic self-setting safety of the MG08.

A sling could be threaded through a lug on the underside of the water jacket and a slot cut in the left side of the butt. A folding auxiliary anti-aircraft backsight was often fitted above the receiver. The short 100-rd belt was generally preferred to the 250-rd type, as it could be carried in a *Patronenkasten* (drum-like box) attached to a bracket on the right side of the receiver beneath the feed way.

Manufacturers included the government factory in Erfurt; Rheinische Metallwaaren- und Maschinenfabrik (Rh.M & MF) in Sömmerda; Siemens & Halske (S & H) in Berlin; and Maschinenfabrik Augsburg-Nürnberg (MAN). A typical wartime gun was marked 'M.G. 08/15.', 'S. & H.', 'BERLIN' and '1918' in four lines, on top of the receiver beneath the serial number; MAN and the

Erfurt factory used 10,000-gun cyclical serial-numbering systems which reached the 'a' block in 1918; the other contractors seem to have preferred continuous numbering from '1' upwards.

The MG08/15 was a useful addition to the German armoury, but, at nearly 19kg, it was appreciably heavier than the guns it was designed to emulate. It was also handicapped by a fixed barrel and water-cooling, even though its sustained fire capability was much better than that of its air-cooled rivals. The MG08/15 proved to be much more reliable than the Lewis Gun, but quickly gained a reputation for poor accuracy.

7.9mm Gewehr 98

Few mechanical changes were made to the *Gewehre* 98 between 1914 and 1918, despite the introduction of several new cartridges. K-Munition, which appeared in the spring of 1915, was loaded with a *Spitzgeschoss mit Kern* (steel-cored armour-piercing bullet). Ammunition of this type was reserved for use in machine-guns and *Scharfschützen-Gewehre* 98, as it was expensive to make.

On 19 November 1915, a new butt fixture consisting of two washers connected by a short hollow tube (to protect the nose of the firing pin during dismantling) replaced the markings disc attached to the right side of the butt. A finger groove appeared in the fore-end at about the same time.

Several modified sights were developed as needs arose. There was an auxiliary *Hilfskorn* or front sight blade (*see* Part Three: '1914'); an anti-aircraft 'lead' sight attachment; three different front sights of varying height, to correct the shooting of individual guns; and a special *Leuchtvisier* (night sight) with luminous radium inserts to improve accuracy in twilight or dusk. There was even a *Gewehr mit Spiegelkolben*, which was a *Gewehr* 98 with a special hinged stock and a periscope mirror sight allowing firing to be undertaken without exposing the operator to danger. Few of these developments had any real significance, and even the luminous night sights – which the troops had been anxious to receive – were abandoned after a short trial.

The Mondragon FSK15, with the drum or 'snail' extension magazine. Ian Hogg

7mm Flieger-Selbstladekarabiner Modell 1915 (FSK15)

The work of a Mexican Army officer, Manuel Mondragon, this gas-operated auto-loader originated in the 1890s, though many years passed before it was patented in the USA in August 1904.

The gun was adopted by the Mexican Army as the *Fusil Automatico de 7mm 'Porfirio Diaz', Modelo de 1908*, and a contract for 4,000 guns was placed in Switzerland. The barrels of the prototypes were 750mm long, but were cut down on series-production guns to save weight. Four hundred Mondragons

7mm *Flieger-Selbstladekarabiner Modell* 1915 (FSK15)	
Synonym:	Mondragon automatic rifle M1915
Adoption date:	2 December 1915
Length of gun:	1,150mm (45.28in)
Weight of gun:	4.31kg (9.5lb) empty
Barrel length:	620mm (24.4in)
Chambering:	7 × 57mm, rimless
Rifling type:	four-groove, concentric
Depth of grooves:	0.13mm (0.005in)
Width of grooves:	3.95mm (0.156in)
Pitch of rifling:	one turn in 220mm (8.65in), RH
Magazine:	detachable box
Loading system:	single rounds
Magazine capacity:	8 or 30 rounds
Front sight:	open barleycorn
Backsight:	tangent-leaf type
Backsight setting:	*minimum* 200m *maximum* 2,000m
Velocity at 80ft:	660m/sec (2,165ft/sec)
Bullet weight:	9g (139gn)

7mm Flieger-Selbstladekarabiner Modell 1915 (FSK15)

Internal Arrangements

Gas is tapped from the bore into an expansion chamber under the barrel, forcing a piston back against the bolt actuator. As the actuator moves back, it rotates the locking lugs out of engagement with the receiver walls. A slight camming action helps to extract the spent cartridge case, then the mechanism runs back to extract and eject before returning to chamber a new round.

A pivoting claw on the bolt-handle can disconnect the bolt assembly from the recoil spring during the cocking stroke, turning the rifle into a manually operated straight-pull pattern if the gas port has been closed.

External Appearance

The Mondragon is a conventional-looking auto-loader, with a one-piece straight-wrist stock and a hand guard stretching forward from the chamber to a nose-cap from which the gas regulator protrudes. A lug for the unique trowel bayonet lies under the nose-cap, one swivel appears on the barrel band, and another is fixed beneath the butt.

reached Mexico early in 1911, but revolution in May 1911 toppled President Porfirio Diaz and the order was cancelled. About 3,000 remained in the SIG factory until 1915, when they were sold to Germany with six box magazines apiece.

The Mondragons were initially tried in the air, prior to the advent of suitable machine-guns, but were prone to jamming. Trials had shown they were not durable enough to withstand the Western Front, so numbers were given to the German Navy for use by naval airmen, airfield-defence units, the observers' schools and the *Marinekorps Sturmtruppen*. Kiel dockyard had nearly 500 of them in October 1918.

11   1916

AUSTRIA–HUNGARY

Captured Russian Rifles

The Austro-Hungarian forces on the Eastern Front captured large quantities of 7.62 × 54mm 1891-pattern Mosin-Nagants in 1914–15, and also received large numbers taken by the Germans. They included substantial numbers of the short-barrel dragoon rifles, often mistakenly known in Austro-Hungarian service as 'Cossack rifles'. Their sights were graduated for 1891-pattern ogival or 1908-pattern spitzer bullets, but many guns were subsequently converted to chamber the standard 8 × 50mm rimmed cartridge and issued to the Austrian *Landwehr*. Most of the unaltered guns seem to have remained with units serving on the Eastern Front, where captured Russian ammunition was available.

Single-shot bolt-action Berdan rifles have also been reported with Austro-Hungarian sling swivels or loops, 'AZF' marks, and unit markings applied by the *Landwehr* or the *Honvéd*. A few were even converted to fire signal and flare cartridges, mating the original butt and action with a new large-diameter tip-down barrel.

8mm Russisches Repetiergewehr *M91*

About 45,000 Russian rifles were converted to fire the 8 × 50mm rimmed Austro-Hungarian service cartridge by the Artilleriezeug-Fabrik in Vienna or Österreichische Waffenfabriks-Gesellschaft in Steyr, being marked 'AZF' and 'OEWG' respectively. The earliest examples may well have had their bores altered, but tests showed that a forcing cone could simply be cut to connect the new chamber to the original Russian four-groove rifling. This not only allowed the 8mm-calibre bullet to be squeezed down to a bore diameter of 7.62mm, but also testified to the strength of the Mosin-Nagant rifles; chamber pressures would have risen appreciably. Problems may have arisen on active service, but the conversions were confined to second-line use and rarely had to be fired in anger.

It is suspected that new backsight leaves were fitted, as the trajectories of the Russian 7.62mm and Austro-Hungarian 8mm cartridges differed. A new *Stutzenkorn* (front sight) was used; loops were added to the fore-end of the short dragoon rifles (which had sling slots instead of swivels); and the swivels of the infantry rifles were altered to suit Austro-Hungarian practice.

The socket bayonets were a mixture of captured Russian and new patterns, with a distinctive straight attachment slot, made by *Erzeugungsabteilung* IX in the Vienna arsenal. Sheet-zinc scabbards were also made in large numbers.

8mm Maschinengewehr *M7/12*

The advent of aerial warfare persuaded the authorities to adapt the M7/12 for a new airborne role. The earliest guns lacked the water-cooling system, their jackets being slotted to allow the circulation of air. The extemporary nature of the conversions created considerable differences in the style of slotting.

The introduction of the M16 turntable for ground guns allowed far greater elevation, depression and traverse – 40, 40 and 33 degrees respectively – than the standard tripod could offer.

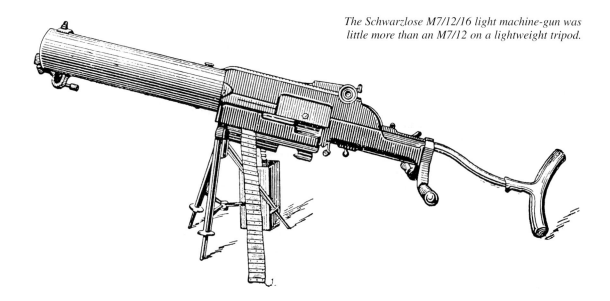

*The Schwarzlose M7/12/16 light machine-gun was
little more than an M7/12 on a lightweight tripod.*

7.92mm Bergmann LMG15

The first light machine-gun of this type was delivered for trials in Austria-Hungary on 3 February
1916, where it performed well enough to encourage the purchase of at least twenty-five guns with
Fokker synchronizer gear. Combat experience
soon revealed the shortcomings of the design (*see*
Part Three: '1915: Germany') and the weapons
had been withdrawn from the *Luftfahrtruppen* by
the autumn. They were subsequently issued to the
balloon detachments for anti-aircraft defence.

8mm Leichtes Maschinengewehr M. 7/12/16

Converting the M7/12 Schwarzlose into an emergency light machine-gun was achieved by substituting a simple bipod for the cumbersome tripod
and attaching a rudimentary butt with a shoulder
pad to the rear of the receiver. Generally known as
the 'M7/12/16', the guns were issued with special
100-round belts. They could fire 1,200 rounds
before their water jackets needed refilling.

The advent of the lightened Schwarzlose
allowed the fourth platoon of each infantry company to be converted into a pair of light machine-
gun sections of two guns apiece, manned by an
officer and thirty-five men.

7.63mm Mauser-Selbstladepistole C. 96

Deliveries of these pistols were made in 1916.
Numbered in the 365,000–410,000 group, they
were standard wartime commercial pistols with
140mm barrels, the new-style safety ('NS'), finely
grooved grips, and backsights graduated from 50m
to 1,000m. Distinguishing marks were confined to
acceptance marks. They should not be confused
with guns sold in Austria-Hungary prior to 1914,
which display dated Budapest, Vienna or Weipert
proof marks.

6.5mm Italienisches Repetiergewehr M91

Large numbers of Italian Mannlicher-Carcano
rifles – and possibly also some *Moschetti* (short
rifles) – were captured on the Southern Front after
1915. Some were reissued to Austro-Hungarian
units without alteration, but a shortage of suitable
6.5mm ammunition and a lack of an indigenous
supplier created problems. These were solved by
altering the chamber dimensions to handle Italian
6.5 × 52mm and Greek 6.5 × 54mm rimless
cartridges interchangeably, as Greek ammunition
was being made by Hirtenberg Patronenfabrik for
the Mannlicher-Schönauers impressed into service
in 1914–15.

The Greek cartridge had a longer case and a slightly larger body, but the neck diameter was the same as the Italian pattern. Firing captured Italian ammunition may have resulted in excessive jamming and case-head separations, but Greek cartridges were more common.

The butts of these adapted Mannlicher-Carcano rifles were marked 'Jt. u. Gr' in 20mm letters. The sights were unchanged, but unique twisted-strip emergency knife bayonets were sometimes issued instead of the standard Italian types.

7.65mm Repetiergewehr *M90/03*

These guns were used in small numbers by the Austro-Hungarian forces – the so-called 'Orient Corps' – fighting alongside the Turks. The units were officially armed with 1895-pattern Mannlichers, but it is assumed that losses of equipment were made good with Turkish weapons. The M90/03 was the charger-loaded M1903 Mauser service rifle, practically identical mechanically to the Gew. 98 (qv). It had a bayonet lug under the nose-cap, a tangent-leaf sight graduated in Arabic, and the *Toughra* of Sultan Abdülhamid II above the chamber (*see* Part Two for further details).

8mm Flugzeug-Maschinengewehr *M7/12/16*

This was a purpose-built jacketless airborne variant of the M7/12, often fitted with special elongated grips with a conventionally guarded trigger fitted to the right grip. The front sight was carried on a rod extending forward from the receiver-cover, supported by a pillar attached to a collar around the barrel. The earliest guns were converted from ground guns and often had a remnant of the jacket at the breech; later examples lacked this identification feature. Altered in the *Fliegerarsenal* in Vienna, the guns had their cyclic rate increased to 550–575rd/min by changing the return spring.

7mm Repetiergewehr *M14*

About 3,000 Chilean-type Mo. 1912 Mausers, left over from pre-World War One orders given to Österreichische Waffenfabriks-Gesellschaft, were issued to Austrian troops in 1916. They were identical with the Mexican-pattern M14 rifle, but had the Chilean Arms – shared with the *Repetierpistole* M12 (qv) – above the chamber instead of the Mexican eagle-and-cactus emblem.

9mm Repetierpistole *M12/16*

Also known as the 'M12/P16', this selective-fire adaptation of the basic Steyr-Hahn design was introduced to bolster firepower on the Southern Front, where the Austro-Hungarians were outnumbered and largely out-gunned by the Italians armed with twin-barrel Villar Perosa machine pistols.

The project was entrusted to Österreichische Waffenfabriks-Gesellschaft at the end of 1915 and fifty adapted M12 pistols were supplied to the *Standschützenbataillon* Innsbruck II in February 1916. Though they worked well enough, a cyclic rate of about 800rd/min and an eight-round magazine proved to be a poor combination. A sixteen-round extension magazine (subsequently standardized) was a feature of the 5,000 guns ordered in 1916.

An inventory taken in the Tirol and on the Isonzo Front revealed that 9,873 M12/16 pistols were still in service at the end of World War One. In addition to the elongated magazine, protruding from the base of the butt, the M12/16 had a radial selector lever on the right side of the frame above the grip. The auto-firing system was eventually granted protection in December 1919, but the papers of Austrian Patent No. 79594 reveal that the application had been made three years previously.

GERMANY

Drum Magazines

The greatest weakness of many semi-automatic assault weapons often lay in the box magazine, which was too small to be useful. One answer was found in the spring-driven *Trommelmagazine* (drum magazines).

Mondragon automatic rifles, adopted in December 1915, were issued for service from the beginning of 1916 and it is probable that the first drum

The Lange Pistole *08 with its drum magazine (TM08).*

units were developed for the FSK15 instead of the LP08. The *Trommelmagazine* had been patented throughout Europe prior to World War One, by Edmund Tatarek and Johann von Benkö, though the design of the Parabellum drum unit is now customarily credited to Friedrich Blum of Budapest. However, although Blum received the relevant German patents in 1915–16, he may have been nothing but a financier promoting an adaptation of the Tatarek-von Benkö magazine.

The drums were complicated and delicate, but occupied much less space beneath a gun than conventional boxes of similar capacity – an important consideration not only in aerial combat, but also in trench warfare. The Mondragon pattern ('TM. für FSK'), which simply slotted into the original feed way, was made in small numbers by Hamburg-Amerikanischen Uhrenfabrik of Schramberg. It increased the capacity of the gun from the eight or ten rounds of the original box magazines to thirty.

The *Trommelmagazin* 1908 (TM08: drum magazine for the 1908-pattern long Parabellum) increased the cartridge capacity from eight to thirty-two, with twenty rounds in the drum body and an extra twelve in the box-like feed way. A spring-driven follower pushed the cartridges around the helix in the drum and up through the pistol butt into the action.

The principal contractor was Gebrüder Bing AG of Nürnberg, Germany's leading manufacturer of tin plate toys, but Allgemeine Elektrizitäts Gesellschaft (AEG) also made some magazines in Berlin. Vereinigte Automaten-Fabriken Pelzer & Companie of Köln (a maker of vending machines in peacetime), was also recruited, but no Köln-made magazines have ever been found.

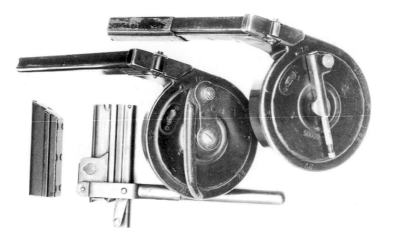

Two Bing-made drum magazines, an older example with a telescoping winding lever and a later folding-lever version. They are shown together with a loading tool and a cover to protect the feed-mouth. Hans Reckendorf

Bing products are marked 'B' (Bing) over 'N' (Nürnberg); AEG's display a quadruple-hexagon trademark; and the prototypes said to have been made by Pelzer & Companie would have borne 'VAF' (Vereinigte Automaten-Fabriken) over 'C' (Cöln).

The earliest Bing magazines had plain bottoms and telescoping winding levers, the magazine-column bracket being retained by two screws. As the cartridges were fired, the winding lever, attached to the magazine-spring, pointed to numbers stamped into the drum body ('32', '27', '22', '17', '12') to show how many rounds were still to be fired.

Service experience soon showed that these bodies were not rigid enough, and a single annular reinforcing rib was pressed into the top surface. The twin-screw magazine-column brace was then superseded by an improved pattern relying on a screw-and-nut fixture. The third-pattern Bing magazine had a folding winding lever, which not only gave better mechanical advantage but was less prone to bending than the sliding design. The fourth pattern was similar, but had a new double concentric-ring reinforcement on the bottom plate.

AEG may have introduced the magazine-body reinforcing ring and screw-and-nut magazine-column bracket, but work ceased after less than

80,000 magazines had been made; Bing made at least ten times as many.

Issued in simple canvas or elaborate leather holdalls, drum magazines were accompanied by light tin plate or sheet-steel dust covers. The first LP08 issues appear to have been made early in 1917, the year in which the official manual, *Anleitung zur langen Pistole 08 mit ansteckbarem*

(Above) *The drum of a Bing-made TM08. Note the 'B/line/N' trademark and the cartridge-remaining numbers.* Weller & Dufty Ltd

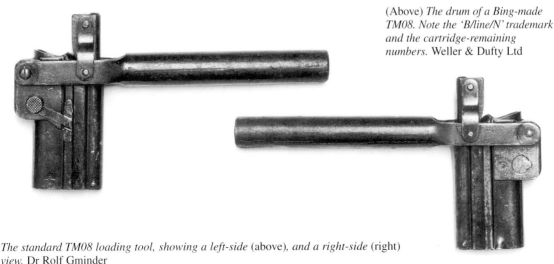

The standard TM08 loading tool, showing a left-side (above), *and a right-side* (right) *view.* Dr Rolf Gminder

Trommelmagazin, stated that the *Kasten für TM08* (magazine box for the TM08) contained five drum magazines, a loading tool and ammunition.

7.9mm *Parabellum*-Maschinengewehre

The failure of the MG08 and even the lightened MG08/15 to provide efficient observers' guns created a new niche for the Parabellum. Guns were normally mounted singly, feeding from a standard Maxim-type fabric belt carried on a reel attached to a bracket on the right side of the gun. A few two-gun mounts were made, but as the Parabellum was not made in 'handed' patterns and could feed only from the right side, the staggered layout was too ungainly to become popular. A lightweight ammunition belt was developed for the air-service Parabellum in 1918, in a successful bid to increase the rate of fire – to 700rd/min – by reducing the weight the feed system was required to lift. Belts of this type had been issued for service, but were by no means universally popular by the time World War One ended.

9mm Pistole *08*

Protected by Luger's German Patent 312,919 of 1 April 1916, a new sear design allowed the firer to cock the mechanism even if the safety-catch was applied. A similar modification was made to long-barrel and Navy guns delivered in 1916–18. The revisions did not affect the official designations, though guns with modified sears were often called *Pistolen* 1908 *mit Abzugsstange neuer Art* (Pistole 1908 with new pattern sear-bar) whilst unaltered guns became *mit Abzugsstange alter Art* (with old-pattern sear bar).

The original truncated 9mm 1908-pattern bullet gave way in mid-1916 to an ogival pattern weighing 7.45–7.97g. This change was made to improve the feed in drum magazines and to prevent Allied propaganda claiming that the flat-headed bullet infringed the Hague Convention.

9mm *Steyr*-Pistole

The Bavarian Army, faced with a shortage of handguns, purchased *Repetierpistolen* M12 ('Steyr-Hahn') from the Österreichische Waffenfabriks-Gesellschaft. Initially, 10,000 were acquired in April 1916, followed by another 6,000 in March 1918, but small-scale deliveries were still being made in October 1918.

Machine-Gun Production

One of the most important features of 1915 had been the inception of the *Hindenburg-Programm*, designed to accelerate the production of war *matériel*. As a consequence, the inventory of machine-guns, which had stood at about 8,000 at the beginning of 1916, rose steadily to 11,000 in July and 16,000 at the end of December 1916; Deutsche Waffen- und Munitionsfabriken's production alone rose from 3,950 in the 1915–16 financial year to 7,350 in the 1916–17 financial year.

On 25 August 1916, the *Kriegsministerium* allowed captured machine-guns to be issued without the permission of higher authority. The publication in November 1916 of *Beute-Maschinengewehre* ('Captured Machine-guns') gave details.

The British provided Colt, Maxim and Vickers patterns, in addition to Lewis light machine-guns. Excepting the 0.30-calibre Colt, these weapons were originally chambered for the standard 0.303 rimmed cartridge (7.7 × 56mm). The Maxims and Vickers guns, which the Germans considered as a group, were issued in three types: *alter Art* (old pattern), *mittlerer Art* (middle pattern) and *neuer Art* (new pattern), weighing 29kg, 19kg and 13.3kg respectively without their mounts. Some of the British guns were subsequently altered to handle the 7.9 × 57mm rimless cartridge, which undoubtedly improved the performance of the Lewis, in particular, by reducing rim-over-rim jamming.

The French Saint-Étienne (Mle 07) and Hotchkiss (Mle 14), each available in *alter Art* and *neuer Art*, were issued unaltered for the original rimmed 8 × 51mm cartridge. The 8mm Fusil Mitrailleur Mle 15 or 'Chauchat' was also available in quantity.

Schwarzlose M07 and M07/12 machine-guns were used in small numbers by German units serving alongside the Austro-Hungarians on the Southern Front.

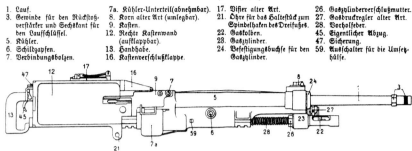

German drawings of the French Mle 07 (Saint-Étienne) machine-gun.

M. G. von rechts gesehen.

1. Lauf.
3. Gewinde für den Rückstoßverstärker und Sechskant für den Laufschlüssel.
5. Kühler.
6. Schildzapfen.
7. Verbindungsbolzen.
7a. Kühler-Unterteil (abnehmbar).
8. Korn alter Art (umlegbar).
9. Kasten.
12. Rechte Kastenwand (aufklappbar).
13. Handhabe.
16. Kastenverschlußklappe.
17. Visier alter Art.
21. Ohre für das Halteftück zum Spindelhaken des Dreifußes.
22. Gaskolben.
23. Gaszylinder.
24. Befestigungsbuchse für den Gaszylinder.
26. Gaszylinderverschlußmutter.
27. Gasdruckregler alter Art.
28. Vorholfeder.
45. Eigentlicher Abzug.
47. Sicherung.
59. Ausschalter für die Umsetzhülse.

M. G. von links gesehen.

1. Lauf.
3. Gewinde für den Rückstoßverstärker und Sechskant für den Laufschlüssel.
4. Sperrhebel für den eingeschraubten Lauf.
5. Kühler.
6. Schildzapfen.
7. Verbindungsbolzen.
7a. Kühler-Unterteil (abnehmbar).
8. Korn alter Art.
9. Kasten.
14. Schloßhebel.
15. Schutzklappe für den Verschlußriegel.
16. Kastenverschlußklappe.
17. Visier alter Art.
21. Ohre für das Halteftück zum Spindelhaken des Dreifußes.
22. Gaskolben.
23. Gaszylinder.
24. Befestigungsbuchse für den Gaszylinder.
26. Gaszylinderverschlußmutter.
27. Gasdruckregler alter Art.
45. Eigentlicher Abzug.
47. Sicherung.
51. Schnellfeuerknopf.
52. Stellhebel zum Feuerregler.
58. Trommelausschalter.
60. Zuführungsplatte.

(Below) *Machine-gunners of a German army unit pose with captured Russian Maxims on Sokolov wheeled mounts, adapted and reissued for service on the Western Front. This is clearly for effect, as none of the ammunition belts contain cartridges.*

Russian Maxim machine-guns had been captured in such large numbers that many were altered in the Spandau factory – *umgeändert für S-Munition* – to chamber the standard 7.9 × 57mm cartridge. According to the Germans, there were two basic patterns: the original of 1905, with bronze jacket and feed block, and a later version of 1910 with steel components. The guns were issued on wheeled Sokolov or tripod mounts, often fitted with new German-made shields.

The Russian 1905- and 1910-type Maxims were captured in large numbers, together with the distinctive wheeled Sokolov mounts. Note the optical-sight bracket on the left side of the receiver. Wallis & Wallis Ltd

(Below) *The parts of the Russian Maxims altered by the Germans for the 7.9mm rimless cartridge. Note the original Russian shields (30, 31) and the Spandau-made German type (34). From* Beute-Maschinengewehre *(1916).*

Erſatzteile zum ruſſ. M. G. geändert für S-Munition.

Die hier aufgeführten Teile müſſen von den Truppenteilen bei der Gewehr-Prüfungs-Kommiſſion angefordert werden (außer Poſ. 33 und 34).

1. Dampfrohr mit Sperrohr.
2. Dampfrohrverſchlußſchraube.
3. Stopfbuchſe.
4. Überrohr.
5. Füll- und Ablaßſchraube.
6. Tülle zum Aufſchieben des Dampfablaßſchlauches.
7. Federbolzen zur Verbindung des M. G. mit der Richtvorrichtung.
8. Verbindungsbolzen für M. G. und Lafette. (Dreifuß.)
9. Verbindungsbolzen für M. G. und Lafette. (Sokolow.)
10. Handhabebolzen.
11. Handhabe und zugehörige Teile. (Bronze oder Stahl.)
11 a. Schraubdeckel mit Pinſel.
11 b. Ölrohr.
11 c. Handgriff.
12. Rechter Verſchlußſchieber (vollſt.).
13. Linker Verſchlußſchieber.
14. Abzugſtange für Bronze- oder Stahlhandhabe.
15. Ausſtoßrohrteile.
16. Kaſtendeckel (ohne Viſierſtange).
17. Deckelriegel.
18. Deckelriegelfeder mit Buchſe.
19. Deckelache mit Ring und Splint.
20. Zuführer und zugehörige Teile. (Bronze.)
21. Zuführer und zugehörige Teile. (Stahl.)
22. Federgehäuſe.
23. Schloßhebel.
24. Kettenhaſpel mit Kette.
25. Schloßkurbel.
26. Kurbelbolzen.
27. Schloßfuß.
28. Linke Gleitwand.
29. Rechte Gleitwand.
30. Schutzſchild (bei Dreifußlafette).
31. Schutzſchild (bei Sokolowlafette).
32. Mantelkopfſchild.
33. Erſatzſchutzſchild (für Dreifuß- und Sokolowlafette).
34. Schildhalter für Sokolomlafette. (Sind mit anzufordern, wenn das Erſatzſchutzſchild zur Sokolomlafette verwendet werden ſoll.)

Gleitvorrichtung.

Bei Art. Werkſtatt Spandau anfordern.

Bei Anforderung iſt unbedingte Angabe erforderlich, ob das M. G. Bronze- oder Stahlmantel beſitzt und ob es auf Dreifuß- oder Sokolowlafette gelagert iſt.

These naval machine-gun teams, pictured in Flanders in 1917, armed with what appears to be a mixture of captured Belgian and German Maxim guns, rely on carts pulled by dog teams.

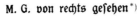

M. G. von rechts gesehen*)

1. Schulterkolben.
2. Kasten.
2b. Trommelauffsteckzapfen.
3. Mantel.
3a. Mantelvorderteil.
4. Vorderer Mantelring mit Korn.

8. Gasregulierschraube.
9. Stellhebel zur Gasregulierschraube.
10. Kühler.
16. Sicherungsschiene
19. Zubringer.
22. Kastendeckel.

27. Visier.
28. Zugfedergehäuse.
33. Mittelachse zur Gewindebuchse.
35. Griffstück zum Abzug.
37. Abzug.
39. Kolbensperrstück.

German drawings of the British Lewis light machine-gun.

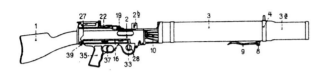

M. G. von links gesehen.

1. Schulterkolben.
2. Kasten.
2b. Trommelauffsteckzapfen.
3. Mantel.
3a. Mantelvorderteil.
4. Vorderer Mantelring mit Korn.

8. Gasregulierschraube.
9. Stellhebel zur Gasregulierschraube.
10. Kühler.
15. Ladeknopf.
16. Sicherungsschiene.
19. Zubringer.

22. Kastendeckel.
27. Visier.
28. Zugfedergehäuse.
35. Griffstück zum Abzug.
37. Abzug.
39. Kolbensperrstück.

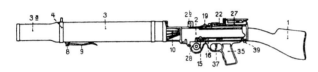

*) Die unterstrichenen Teile sind bei der Umänderung für deutsche Munition abgeändert.

Alterations varied from gun to gun. The Russian Maxims required the most changes, ranging from new barrels, feed blocks and back plates (complete with trigger and spade grips) to water-tubes for their barrel jackets. The Vickers Gun was easier to adapt, requiring only a new backsight assembly, a feed-cover, a feed block, two feed claws, a belt pawl, a carrier block, a cartridge retainer, a spring plate, and breech-block cam plates. The barrel, barrel-bush and muzzle cone also needed changing.

The Lewis Gun required a few machining revisions in the receiver, a new barrel, a new gas-regulator plug, new extractors, a new feed plate, altered sights, a new magazine top, and a new stop-screw for the magazine spring.

Machine-gun sections, subsequently enlarged to squadrons, were attached to the German Army's cavalry regiments early in 1916. Machine-gun companies were then attached to cyclist battalions, with six guns mounted individually on three-tonne lorries.

The increased availability of machine-guns allowed issue, which had become fragmented, to be standardized in August 1916 on the basis of a six-gun regimental machine-gun company organized in three sections of two guns. The creation of three machine-gun companies per regiment allowed a company to be attached to each infantry battalion. The first *Maschinengewehr-Scharfschützen-Abteilungen* (machine-gun marksman detachments) were also formed at this time. Operating independently under the command of General Headquarters, their personnel wore a special machine-gun badge on the left sleeve.

Rifle Production

Wartime production of the Mauser rifles was accelerated as quickly as practicable, as more and more sub-contractors were recruited. Virtually every uncommitted metalworking company in Thüringen – traditionally the centre of the German firearms industry – was ordered to make parts for the *Gewehr* 98 or Maxim machine-guns, supplying large numbers of extractors, ejectors, bolt-stops, magazine floor plates, trigger-guard bows and similar parts to the government arsenals.

Teams of experienced inspectors assembled the component parts to make rifles known as *Stern Gewehre* (star rifles), owing to the large ★ struck into the top of each chamber. Many of their parts would have failed pre-World War One inspection, but the standards were altered to accept virtually anything that was structurally sound and operated satisfactorily. The penalty was simply that complete interchangeability could no longer be guaranteed.

7.9mm Gewehr 98

Until 1916, all rifles were stocked in walnut. However, supplies of properly seasoned wood soon ran low as production of weapons increased: stock

A sailor of the Marinekorps, *guarding the German seaward flank in Flanders, surveys an expanse of coastal estuary. He is armed with a* Gewehr 98.

blanks had to be seasoned for at least three years before they could be used. Accelerated drying processes were tried without lasting success, so beech and birch substitutes were authorized on 7 June 1916.

A minor alteration was made in this period to the front action-retaining screw, which ran up through the front end of the trigger-guard bow extension/magazine floor plate assembly. Only one lock-screw position was provided, thus simplifying manufacture.

Aircraft Machine-Guns

The first aircraft guns were 1908-type Maxims stripped of all superfluous components. The water-cooling system was removed – it was irrelevant in the air, even at the 100-kt speeds of the first fighters – and changes were made to accept synchronizing gear.

The first German warplane to be fitted with a synchronizer was a Fokker E.I *Eindecker* (*see* Part Three, '1915: Germany'), and all the conversion work prior to December 1916 was undertaken in the Fokker factory in Schwerin. The basis of the *Gestängesteuerung* interrupter was a push-rod driven from the engine, but combat experience showed this to be inefficient at high speed. It was not unknown for pilots to shoot blades off their propellers in cold weather.

The push-rod system was replaced in the summer of 1916 by a cable-driven *Zentralsteuerung* pattern designed by Lübbe, Heber and Leimberger. This was much more acceptable and, in December 1916, all work on synchronizing gear and associated machine-guns was transferred to Flugzeug Waffenfabrik GmbH, a newly established Fokker subsidiary in Reinickendorf, then on the outskirts of Berlin. This agency subsequently became the major supplier of guns and equipment to the German aircraft manufacturers.

The weight of the perfected LMG08/15 – there is still some doubt whether the 'L' prefix represents *leichte* (light) or *luftgekuhlt* (air-cooled) – was pared down to 12.5kg without its control gear. The barrel casing was slotted sheet-steel, special charging systems and mounts being developed as necessary. Air-cooled Maxims were occasionally mounted as observers' guns, with pistol-grips and special short butts instead of specialized control equipment, but the lighter and handier Parabellum was preferred.

7.9mm Maschinengewehr Modell *1916* (MG16)

Amalgamating the light and heavy Maxims created this *Einheitsmaschinengewehr* (single-purpose gun). Made by the Spandau rifle factory in small numbers, the resulting MG16 was basically an MG08/15 with MG08-type spade grips and back plate-mounted trigger. It was instantly recognizable by the cutaway receiver and the large folding backsight.

The MG16 was issued with a simplified tripod known as the *Dreifuss* 16, which could also be fitted to the MG08 if an appropriate adaptor was available. The tripod retained the elevating system of the *Schlitten* 08, saving the costs of re-tooling, but was considerably lighter and much easier to make.

Tests showed that the combination of the new gun and new mount was satisfactory. Although the MG16 possessed neither the outstanding sustained-fire capability of the MG08 nor the portability of the MG08/15, this seemed a small price to pay for the virtues of a universal machine-gun.

Problems arose when attempts were made to translate ideas of universality into series production.

World War One was by then into its third year, and, with the advent of mechanical synchronization in air warfare, the need for machine-guns had accelerated so greatly that additional contractors such as Siemens & Halske and Maschinenfabrik Augsburg-Nürnberg had been recruited. However, although basic dimensioned drawings were supplied to each manufacturer, the pressing necessity to produce guns that worked prevailed at the expense of interchangeability.

Realizing that insistence on the use of master drawings would force manufacturers to conform with standard gauges, deviating neither from prescribed tolerances nor material specifications, the authorities reluctantly accepted that the risk of disrupting efficient production of the MG08 and MG08/15 outweighed the advantages of introducing the MG16. The new machine-gun was thus abandoned after only a few semi-experimental examples had been made, though the versatile *Dreifuss* 16 was ordered into immediate production as it was far easier to make than the *Schlitten* 08. Capable of fitting the MG08/15 without modification, the tripod could be issued with the MG08 simply by providing an adaptor.

9mm Pistole *04*

On 29 August 1916, the *Reichs-Marine-Amt* placed a new 8,000-gun order with Deutsche Waffen- und Munitionsfabriken. The wartime guns were similar to

The 1916-type Pistole *04, with a short frame.* Weller & Dufty Ltd

their predecessors, except that they had short frames and lacked the grip-safety mechanism. Their chambers were usually dated '1916' or '1917', though surviving Kiel inventories reveal that guns were still being delivered at the end of World War One.

TURKEY

German Aid

The Allied invasion of the Dardanelles in 1915 brought Turkey into World War One with a vengeance. Soon, German advisers were reporting that the Turkish Army would need to be bolstered with men and equipment if the Allied attacks were to be rebuffed. The pro-Central Powers stance of Bulgaria and the occupation of large parts of Serbia

by German and Austro-Hungarian forces encouraged the dispatch of trainloads of equipment to the Bosphorus. The first consignment left Berlin on 19 January 1916; by the end of World War One, 230,000 rifles, 22,000 handguns and nearly 2,000 machine-guns had been sent to Turkey.

7.9mm Gewehr 88/S, 88/05 and 88/14
Large numbers of these obsolescent rifles were sent to Turkey in 1916–17. Most of them had the backsight graduations replaced with Arabic equivalents, and will be found with small crescent inspectors' marks struck over or substituted for the original German ones. So many guns were left in Turkey by the Germans at the end of World War One that survivors were reconditioned during the 1920s and 1930s ('Model 88/35'), receiving a new pistol-grip stock and many other refinements.

Conditions on the Western Front could be appalling after heavy rain. This picture, taken behind Allied lines in 1917, shows the conditions in which small arms were still expected to function.

12 1917

AUSTRIA–HUNGARY

8mm Maschinengewehr *M7/16A*

Also known as the '7/12/16A', this machine-gun had an extra-strong recoil spring and an additional muzzle booster to increase cyclic rate progressively to 880rd/min by the end of World War One. Conversion kits, made by Jacob Löhner & Cie in the Floridsdorf district of Vienna, could be added in the field to otherwise standard M07/12/16 guns.

6.5mm Japanischen Repetiergewehr *M97*

Limited use was made of Japanese Arisaka rifles captured from the Russians on the Eastern Front, but it is not known whether they were all genuinely *Meiji* 30th Year Type guns (the so-called 'hook safety' pattern) or a mixture of these and the later 38th Year patterns. A *Marsch-Bataillon* of *Infanterie*-Regiment Nr 14 was equipped with Arisakas in 1918, but it is assumed that the total quantities involved were small.

Chambered for a 6.5mm semi-rim cartridge, the Arisaka was built on a modified Mauser action. Stocked conventionally, though often with a butt made in two parts, it had a 2,000m leaf-pattern backsight and accepted a sword bayonet with a hooked quillon.

GERMANY

7.9mm Gewehr *98 and* Karabiner *98 AZ*

Gunstocks made of *Ahornholz* (European maple) were approved on 10 January 1917, though, until the desperate days of 1918, these rifles were supposedly restricted to second-line units. Their butts bore a large 'A'. Elm stocks and hand guards were made in small quantities, and laminated patterns were developed experimentally. Some guns also had separate butt-toes, which were attached by gluing and a dovetail joint – in the Japanese manner – to make use of stock blanks that would otherwise have been rejected.

A stamped-steel bolt cover was added to some guns to prevent trench mud getting into the bolt mechanism. A light sheet-steel body fixed over the receiver, retained by a clip under the bolt-handle shank, reciprocated with the bolt. A rod extending forward from the front left side of the cover slid in a collar attached around the barrel and fore-end by a powerful spring clip. Bolt covers of this type are usually marked 'W–CO/D R P', but the manufacturer has yet to be identified and it is assumed that they had no official status.

Auxiliary twenty-round magazines were developed for use in the trenches, but never became common as changes were required in the magazine floor plate.

Production peaked in 1917, when several of the major manufacturers – for example, Erfurt arsenal – used the entire 'a'–'z' set of serial-number suffix letters and began 'aa'–'zz', indicating that more than 270,000 rifles had been made in the previous twelve months.

The Mauser continued to serve virtually all the front-line units in the German Army, and a report from Flanders in July 1917 indicated that Gew. 98 were also being carried by the *Marine-* and *Matrosen-Infanterie*, the coast and anti-aircraft artillery, and the *Marine-Pionier-Kompagnien*;

Members of the seventh company of Bavarian Landwehr-Infanterie-Regiment Nr 12 *pose with their* Gewehre 98 *in August 1916. A few of the guns are fitted with twenty-round extension magazines.*

naval issue of the Kar. 98 AZ was restricted at this time to the field- and foot-artillerymen, airmen and balloonists.

7.9mm and 9mm Ammunition

Nickel-plating of bullets was replaced by tombac alloy to conserve raw material, and, from 8 March 1917, the primer cup could be made from an alloy of 90 per cent lead and 10 per cent zinc.

Experimental steel-case cartridges were also made in this period, and a broad selection of special-purpose ammunition – tracer, armour-piercing and ranging rounds among them – was introduced for the machine-guns.

7.9mm Maschinengewehr Modell *1914/17 (MG14/17)*

This was the LMG14 or Parabellum (qv) altered so that it was easier to manoeuvre in the slipstream of aircraft, which were becoming increasingly faster. The barrel was shortened, the diameter of the barrel casing was reduced to a minimum, and two brackets were attached to the top of the receiver to accept an optical sight.

Substitute Pistols

Low-power guns, often of inferior design and construction, were usually issued to men of the Train and lines-of-communication units to free more

combat-worthy pistols for combat duties. *Preise für Pistolen und Revolver*, produced for the *Kriegsministerium* in August 1917, listed a wide selection of acceptable *Behelfspistolen*.

Konteradmiral *Wedding and his aides survey a map. The aide on the left has a small pistol – probably a 7.65mm example – in a belt holster.*

They included the Mauser C96: 7.63mm or 9mm Mauser with stock, 7.63mm without stock. This well-known design has already been described (*see* Part Two, Germany: 'Handguns'), as has the 7.65mm Frommer 'Stop', or Pisztoly 12M (*see* Part Two: 'Austria–Hungary').

Among the guns seized after the invasion of Belgium were the 9mm Pieper; the 9mm 'Bayard, large' (Bergmann-Bayard); the 9mm Bayard, small; the 9mm 'small' FN-Browning (Mle 10); a 9mm 'large' FN-Browning (Mle 03) with an optional stock; a 7.65mm Pieper; a 7.65mm Bayard; and 6.35mm *Lütticher Pistolen* ('Liége pistols') of various patterns.

The German guns included, surprisingly, the short-lived 9mm Walther *Modell* 6. Most of the 7.65mm designs – Beholla, Dreyse, Jäger, Langenhan, Mauser, Menta, Sauer and Walther – are described in the previous chapter. There were also 7.65mm pistols made by Meffert of Suhl, which, rather oddly, the list describes as 'Walther and Dreyse' types without explanation.

Guns that were accepted officially bore crowned-letter inspectors' marks, though this was obviously not as true of the guns that had been purchased privately by individual officers.

7.9mm Gewehr Modell *1898/17* (Gew. 98/17)

This was a *Gewehr-Prüfungs-Kommission* inspired revision of the standard Mauser infantry rifle, designed as a better *Nahkampfwaffe* (close-combat weapon). The barrel was altered from conical to cylindrical – to save weight and make production easier – and the minimum sighting distance was only 100m. A tangent-leaf backsight replaced the Lange Visier; an improved stamped-steel bolt cover was developed and adapted to permit quick-loading from the chargers; an altered magazine follower held the bolt open when the last cartridge had been fired and extracted; and the upper part of the trigger was ribbed to improve grip. It is assumed that most of the other features remained unaltered, but no surviving gun is known to exist.

The FN-Browning Mle 1900 pistol was among the guns seized after the invasion of Belgium in 1914. FN Herstal SA

The 6.35mm FN-Browning pocket pistol was used by German officers as a self-defence weapon. FN Herstal SA

The *Gewehr-Prüfungs-Kommission* ordered 5,000 *Gewehre* 98/17 from Simson & Cie of Suhl in 1917, the first guns being delivered in March 1918. They were to have been issued for large-scale trials later in the year, but World War One ended before anything could be done.

Fully Automatic Pistols

The German Army, like all pre-1914 armies, had once been keen on the pistol-carbine concept. Cavalry and mounted artillerymen often carried bolt-action magazine carbines, but these were comparatively clumsy and often unnecessarily powerful.

A solitary *Maschinenpistole* 08 was demonstrated to the *Gewehr-Prüfungs-Kommission* in December 1917. The otherwise standard *Pistole* 08 action had been modified to fire automatically from drum magazines. Groups were obtained measuring 43 × 55cm at 100m, 100 × 105cm at 200m, and 168cm square at 300m; 50 per cent dispersion at 100m had been 10 × 15cm.

The *Maschinenpistole* 08 was apparently intended to be issued with a special barrel sleeve and the mounting pivot of the Parabellum light machine-gun, and had a rifle-type stock.

Fully automatic pistols emptied their magazines almost instantly, heating the chamber to a point where a round could ignite without the assistance of the firing pin (known as 'cooking off'). As light weight also made the conversions difficult to control when firing automatically, work was soon abandoned.

Machine-Gun Production

Deutsche Waffen- und Munitionsfabriken alone made more than 20,000 machine-guns during the 1917–18 financial year, allowing the allocation of guns to each infantry machine-gun company to increase from six to twelve. The introduction of the MG08/15 allowed the previous light machine-guns – Bergmanns and Madsens – to be withdrawn and, by the end of 1917, at least three MG08/15 machine-guns had been issued to each infantry company. The first *Flugabwehr-Maschinengewehr-Abteilungen* ('Flamga': anti-aircraft machine-gun detachments) were created during this period, each comprising three companies with six-man crews serving twelve MG08.

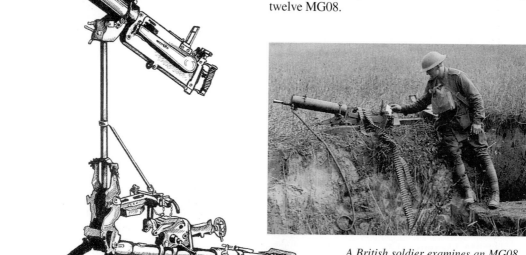

A British soldier examines an MG08 on an improvised static mount.

An MG08 mounted on the 1916-pattern tripod, with an anti-aircraft adaptor. Firepower International

13 1918

AUSTRIA–HUNGARY

7.92mm *Gebauer*-Maschinengewehr

A Czech engineer, Ferencz Gebauer, developed this interesting twin-barrel weapon in 1917. Sometimes known as the '*Gebauer-Weich-Motor-Maschinengewehr*' (Weich being the financier), it was driven from the engine crankshaft by layshafts, bevel gears and a friction clutch. The maximum fire-rate, which depending on engine speed, reached 1,200rd/min. Ammunition was delivered by geared drive to ensure accurate indexing, and the breech was a slam-fire type.

Three prototypes made by Sollux of Vienna were tested so successfully in June 1918 that 100 were ordered immediately. The guns had all been completed by September 1918, but World War One ended before they could be installed in aircraft.

Wartime Production

By the end of World War One, the Austro-Hungarian factories had made a surprisingly large quantity of weapons. These included 2,891,138 Mannlicher rifles and 702,337 carbines or short rifles made by the Steyr and (to a lesser extent) Budapest factories from 1 August 1914 until 1 October 1918. The inventory of machine-guns on 30 October 1918 stood at 43,777.

GERMANY

Few significant innovations occurred in World War One, for although the introduction of the Bergmann *Maschinenpistolen* 18 ultimately called the tactical role of the infantry rifle into question, there were too few of them to make any notable impact before the Armistice.

Machine-Guns

The German armed forces possessed machine-guns in abundance in January 1918, when the inventory stood at about 32,000 heavy and 37,000 light guns. The principal service weapons were water-cooled MG08 and MG08/15 Maxims, but there were also many air-cooled Maxims and Parabellums plus smaller numbers of the Bergmann MG15 and a tiny number of water-cooled Dreyse ground guns.

The establishment of a machine-gun company had finally stabilized – on paper, if not always in the field – as four officers, 133 NCOs and men, six heavy guns, six handcarts, nine vehicles and twenty horses.

13mm Tank-Gewehr Modell *1918* (*T-Gewehr*)

This enormous gun, nicknamed *Elefantenbüchse* (Elephant Rifle), was introduced to counter the first tanks. Development work began on the gun and cartridge at the end of November 1917 – ammunition being credited jointly to Mauser and Polte-Werke of Magdeburg – and the first prototype was demonstrated on 19 January 1918. Full-scale production was authorized just two days later and, remarkably, deliveries were underway by the end of May 1918. Mauser had completed 15,800 T-*Gewehre* by the end of World War One, though by no means all were actually issued. Two T-*Gewehre* were initially issued to each infantry regiment, one

The 13mm Tank-Gewehr *of 1918.* Ian Hogg

(Below) The 13mm Tank-Gewehr *cartridge.*

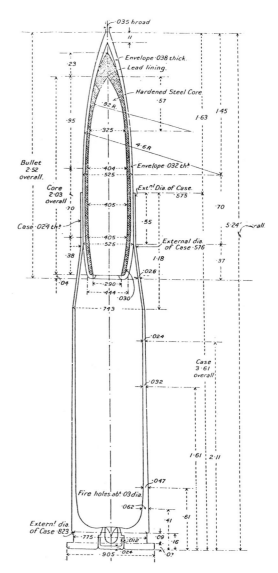

13mm *Tank-Gewehr Modell* 1918 (*T-Gewehr*)

Synonym:	Mauser anti-tank rifle M1918
Adoption date:	January 1918
Length:	1,702mm (67in)
Weight:	17.3kg (38.14lb) with bipod
Weight of mount:	0.81kg (1.79lb)
Barrel length:	870mm (34.25in)
Chambering:	13 × 91mm, semi-rim
Rifling type:	eight-groove, concentric
Depth of grooves:	0.35mm (0.009in)
Width of grooves:	3.5mm (0.138in)
Pitch of rifling:	one turn in 407mm (16.02in)
Loading system:	single rounds
Front sight:	open barleycorn
Backsight:	tangent-leaf type
Backsight setting:	*minimum* 100m
	maximum 500m
Muzzle velocity:	780m/sec (2,560ft/sec)
Bullet weight:	51.5g (795gn)

being kept back from the front-line for training purposes, though the ultimate goal was apparently to provide three guns for each unit.

The T-*Gewehr* was built on an enlarged Mauser action, with two lugs on the bolt head and two in front of the bolt-handle, which locked vertically behind the chamber and vertically within the receiver-bridge respectively. The rifle had a straight-wrist half stock of ash or elm, with a separate pistol-grip behind the trigger and a folding bipod on the tip of the fore-end.

The large brass bottlenecked cartridge case was loaded with a boat-tailed spitzer bullet; jacketed in copper alloy, the bullet had a steel core set into a lead/antimony liner. Loaded rounds were about 132.5mm long and weighed 116g. Tests showed that the projectile could penetrate 30mm of face-hardened steel at 50m, and could still pierce 25mm at 250m as long as the core remained intact. T-*Gewehre* were used by two-man teams, the firer carrying the gun, twelve rounds of ammunition and the tools, whereas his helper had forty rounds in belt pouches and seventy-two in a separate box. Both men carried *Pistolen* 08 and bayonets for short-range defence.

Shortly before World War One ended, a few T-*Gewehre* were made experimentally with detachable five-round box magazines and spring-loaded shoulder plates to absorb some of the fearsome recoil. The trigger-guard bow was extended forward to reach the magazine.

New Ammunition

The heavy *schweres Spitzgeschoss* ('sS') bullet was introduced on 28 July 1918 for use in heavy machine-guns, but was not issued to riflemen during World War One. The boat-tailed projectile was 35.3mm long and weighed about 12.8g, substantially greater than the 'S' pattern. Muzzle velocity was reduced to about 785m/sec in a Gew. 98 and 755m/sec in a Maxim machine-gun.

9mm Pistole 04

On 1 February 1918, according to Hans Reckendorf in his book *Die Handwaffen der Koeniglich Preussischen und der Kaiserlichen Marine* (1983), the *Marinekorps* inventory stood at 10,728 pistols. However, it is not clear how many of these were Parabellums nor from which sources they had come.

'Spandau Lugers'

Mystery still surrounds the *Pistolen* 08 said to have been made in the Spandau factory in 1918.

Manufacture of the Parabellum was so complicated that only four sets of machinery were ever made, and their movements and ultimate fate are well documented. However, as Spandau was the headquarters of the *Revisions-Commission*, a few pistols may have been assembled from parts taken from the guns that had failed proof.

Most of the so-called 'Spandau' guns incorporated receivers made by Deutsche Waffen- und Munitionsfabriken, which can be identified by the 'crown/T', 'crown/S', 'crown/S' inspectors' marks on the right side. Some guns have the 'crown/RC' mark of the *Revisions-Commission*; most also have additional inspectors' marks on the receivers, suggesting that they may have undergone the second and third stages of proof for a second time.

It seems likely that the Spandau *Pistolen* 08 were made in desperation in the early part of 1918, when every available serviceable gun was required for the last great offensives mounted by the Germans during World War One.

9mm Lange Pistolen 08

Many long-barrel guns were withdrawn from the artillery and the assault units in 1918, for issue to the crews of gunboats and inshore minesweepers. A few guns of this type have been reported with German Navy 'crown/M' marks; unit markings had been abandoned by this time.

7.9mm Maschinengewehr Modell 1908/18 (LMG08/18)

Introduced in the last year of World War One as an expedient, this air-cooled Maxim had its origins in air service. Little more than an MG08/15 with the barrel protected by a small-diameter slotted sheet-steel sleeve, the gun had a simple carrying handle attached to the jacket near the breech and a front sight on top of a post.

The air-cooled LMG08/18 Maxim had certain advantages compared with the water-cooled MG08/15, as it did not freeze in winter and was about 1kg lighter. The absence of 3lt of coolant saved another 3kg, and the omission of the steam tube saved 250g. However, an overheating barrel could only be removed backwards through the receiver after the shoulder stock and some internal components had been detached. This restricted the sustained-fire capability of the MG08/18 in a ground rôle to only a few hundred rounds.

7.9mm Gast-Maschinengewehr *M1917*

Based on patents granted in January 1916 and February 1917 to Carl Gast, this machine-gun was made in small numbers by Vorwerk & Cie of Barmen. Three thousand were ordered in the autumn of 1917, and assembly continued into 1919 before work finally stopped.

Despite its archaic appearance, the Gast was an interesting way of increasing the chances of a hit in aerial combat. Realizing that the cyclic rate needed to be higher than even a perfected Parabellum equipped with lightweight ammunition belts, Gast elected to use two barrels within separate slotted casings.

The recoil of one bolt unit was balanced by the counter-recoil or 'run-out' of its companion, locked by rotating collars that released the bolt only in the open position. The coupled action allowed Gast to replace the massive mainsprings with small buffers behind each bolt.

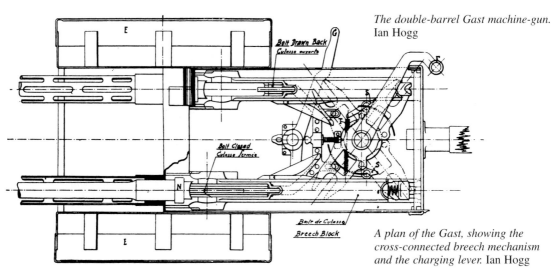

The double-barrel Gast machine-gun.
Ian Hogg

A plan of the Gast, showing the cross-connected breech mechanism and the charging lever. Ian Hogg

Two spring-driven drum magazines were attached vertically, one on each side of the receiver. Each held 180 rounds. The sights and shoulder stock were placed centrally, and a charging handle projected from the right side of the breech. Cyclic rate has been reported as 1,200, 1,600 or 1,800 rd/min, but the gun weighed only about 40lb empty.

Tested, but predictably rejected as an infantry weapon, the Gast was seen as an ideal gun for observers. However, the bluff profile of the receiver hindered manoeuvrability in a slipstream and a reliance on 'handed' magazines demanded vigilance on the part of the gunner. As a breakdown on one side automatically disabled the other, the Gast had its fair share of problems. The Allies reportedly destroyed 1,314 Gast machine-guns at the end of World War One, but few had seen combat and it is likely that less than 500 had been assembled by the Armistice. Guns were still being discovered long after World War One had ended, including a cache found in Königsberg in the early 1920s.

As the motor-driven prototypes had barely left the drawing board by 1918, excepting the Austro-Hungarian Gebauer, the Gast represents one of the few successful attempts made to provide truly rapid fire during World War One. Had the design been refined into a more compact belt-fed package, as the Russians subsequently did with coupled guns developed in the 1920s, the Gast might have exerted greater influence on machine-gun design than it did.

7.9mm *Mauser*-Gewehr *18*

Officially known as the Mauser *Schützengraben-und Nahkampfgewehr* 18 (Mauser trench-shooting and close-combat rifle 18), this was a private development undertaken by Waffenfabrik Mauser AG of Oberndorf shortly before the end of World War One. Several improvements were made in the basic design of the *Gewehr* 98, the most important being the addition of a detachable box magazine for five, ten or twenty-five rounds.

The magazines had stamped vertically-ribbed sides and were retained by a spring catch in the front of the trigger-guard bow; they could be loaded through the open action from the standard

chargers or filled with individual cartridges before being inserted in the feed way. A separate mechanical hold-open, operated by the bolt follower, held the bolt back when the last spent case had been ejected; and a linkage between the trigger and the bolt catch prevented removal of magazines unless the bolt was open.

The mainspring was altered, and the bolt-stop was removed from the rear left side of the receiver to the trigger mechanism. An improved bolt cover was fitted, the pistol-grip was strengthened and the sights were modified. The rifles were only made in prototype form, as the development of the submachine-gun put an end to work on 'trench carbines'.

9mm *Bergmann*-Maschinenpistole *'18,I'* (MP18,I)

Patented by Hugo Schmeisser in December 1917 and April 1918, and made by 'Theodor Bergmann, Waffenabteilung Suhl', this was the first successful

---

9mm Bergmann-*Maschinenpistole* '18,I' (MP18,I)

Synonym:	Bergmann submachine-gun M1918
Adoption date:	none(?)
Length:	815mm (32.09in)
Weight:	4.175kg (9.2lb) empty
Barrel length:	200mm (7.87in)
Chambering:	9 × 19mm, rimmed
Rifling type:	four-groove, concentric
Depth of grooves:	0.125mm (0.049in)
Width of grooves:	2.75mm (0.108in)
Pitch of rifling:	one turn in 250mm (9.84in), RH
Magazine:	spring-driven drum
Magazine capacity:	32 rounds
Loading system:	single rounds
Front sight:	open barleycorn
Backsight:	two-position rocking 'L'
Backsight setting:	minimum 100m maximum 200m
Muzzle velocity:	365m/sec (1,200ft/sec)
Bullet weight:	8g (123gn)
Cyclic rate:	400rd/min

The Bergmann MP18,I, with its drum magazine.

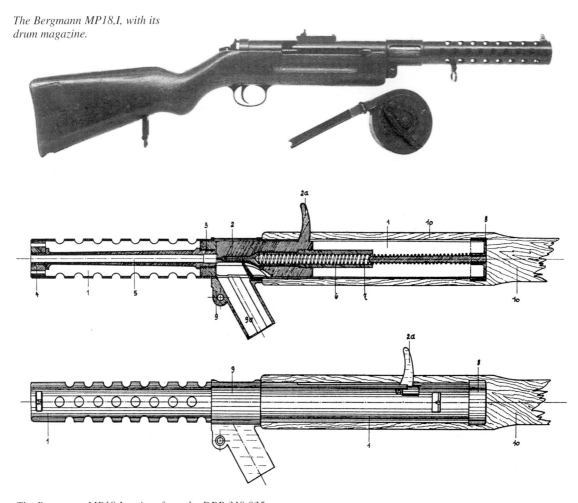

The Bergmann MP18,I action, from the DRP 319,035.

submachine-gun. However, the inflated reputation of the MP18,I rests largely on the unproven supposition that it was used in large numbers during World War One. Even the designation has been questioned. Writing in 'Wieso Maschinenpistole '18,I'?' in the *Deutsches Waffen-Journal* in December 1983, Joachim Görtz suggested not only that '18,I' referred to 'Title 18' provisions in the military budget (which could be shown to govern the replacement of small arms), but also that there was no evidence to show that the MP18,I had ever been officially approved.

Series production had begun by the spring of 1918 – the first manual was issued in April of that year – but claims that 12,500–50,000 guns had been made prior to the Armistice lack authentication. The truth remains unclear, as does the claim that many guns were assembled after hostilities ceased.

The MP18,I had a pistol-grip half stock with a finger groove in the fore-end. Swivels lay beneath the butt and the barrel casing, which was pierced with ventilating holes. The backsight lay on the block above the receiver. The only safety feature (a poor one) was provided by pulling the charging-handle head back, up and into an auxiliary slot. Pressing the thumb-latch projecting from the stock above the back of the receiver allowed the entire action to pivot down around the lower front of the fore-end.

The mechanism relied solely on the weight of the breech-bolt and the pressure of the return spring to delay the initial backward motion until chamber pressure had dropped to an acceptable level. However, though the firing mechanism was a 'slam' type, the MP18,I was heavy enough to enable it to be kept under reasonable control when firing short bursts.

Its greatest weakness was the TM08 (drum magazine), shared with the Parabellum pistol, which forced Schmeisser to provide an angled feed way projecting laterally from the left side of the breech. There was logic in this, because the TM08 was available in large numbers and could be used virtually without modification. Indeed, many TM08 are said to have been withdrawn from pistoleers in 1918 and reissued with a collar-type adaptor around the feed extension to fit the MP18,I. However, the angled feed had never been entirely reliable and the basic design proved to be far more

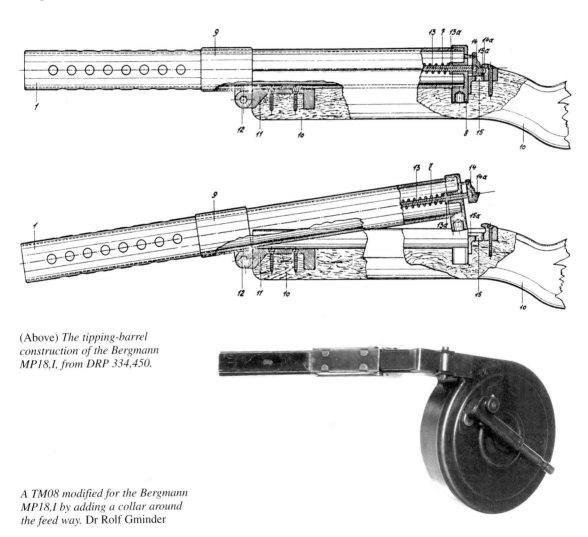

(Above) *The tipping-barrel construction of the Bergmann MP18,I, from DRP 334,450.*

A TM08 modified for the Bergmann MP18,I by adding a collar around the feed way. Dr Rolf Gminder

efficient once it had been adapted to feed from a conventional box magazine.

7.9mm Leichtes Maschinengewehr Modell 1918 (LMG18)

An adaptation of the water-cooled Dreyse machine-gun, perhaps inspired by the introduction of the Browning Automatic Rifle, this air-cooled gun was developed experimentally in the closing stages of World War One. Recoil operation and the pivoting-strut locking system were retained, but the Dreyse fed from a detachable box magazine projecting laterally from the breech. The mainspring was moved to the top of the receiver above the pistol-grip to save space; a shoulder plate was fitted to a tubular stock; and the hammer-type firing mechanism was notably compact.

13mm Maschinengewehr Modell 1918 (MG18 TuF)

The introduction of tanks encouraged the Germans to develop a 13mm cartridge with sufficient penetrating capabilities to defeat primitive armour-plate. The original intention was to provide a machine-gun, but development was protracted and the Mauser T-*Gewehr* – an enlargement of the standard bolt-action rifle – appeared instead. The 13mm MG18 *Tank- und Flieger* (TuF) machine-gun was basically an enlargement of the abortive 8mm *Einheits*-MG16 on a wheeled carriage adapted from the standard 1916-pattern tripod. Prototypes had been tested on Fokker D.VII aircraft by the end of World War One, but few of the 4,000 guns in the course of production were ever completed.

20mm Becker Cannon

This 2cm machine-cannon, made by Stahlwerke Becker (but allegedly designed by the Cönders brothers), proved to be more successful than the 13mm MG18 TuF. The blowback-operated Becker gun is said to have been used in the closing stages of the air war; Allied inspectors subsequently found 362 completed guns, and it has been claimed that 100 or more had already entered service.

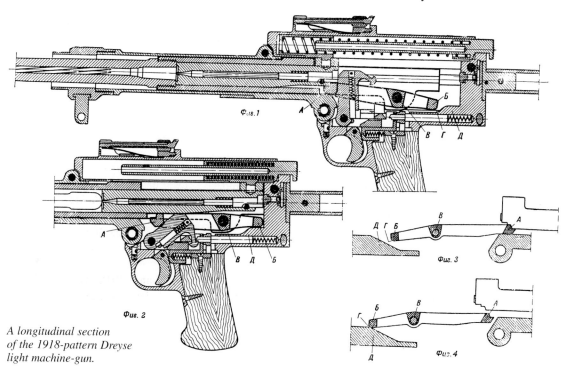

A longitudinal section of the 1918-pattern Dreyse light machine-gun.

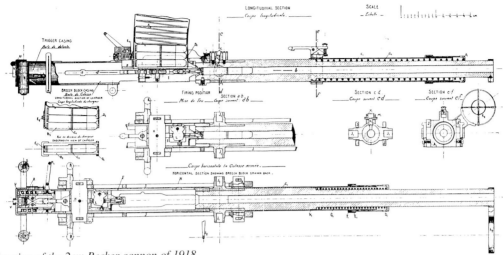

A drawing of the 2cm Becker cannon of 1918.

Though the Becker gun appeared too late in the war to be successful, the basic design was licensed to the Swiss Semag company and ultimately to Oerlikon-Bührle.

Gains and Losses: Weapons Production

Exactly how many machine-guns had been made in Germany prior to 1918 is impossible to determine. Only about 1,600 had been available to the German Army in August 1914, but production of the Maxims, in particular, had grown significantly by the Armistice.

The Inter-Allied Military Control Commission collected 87,950 machine-guns in 1919–20, showing that – allowing for war losses – well over 100,000 had been made.

Between 1 August 1914 and 1 November 1918, Deutsche Waffen- und Munitionsfabriken alone made 930,000 1898-pattern Mauser rifles and short rifles (Gew. 98 and Kar. 98 AZ); 680,000 Parabellum or 'Luger' pistols; 58,000 Maxim and Parabellum machine-guns; 4,000,000,000 small arms cartridges; 22,000,000 shell cases; and 580,000,000 primers in a variety of shapes and sizes.

Losses were also prodigious. Despite acquiring at least 2,000,000 handguns, an inventory dated 1 February 1918 revealed that the field army (including the *Marinekorps*) possessed only 811,109 serviceable pistols, with an additional 228,032 on home service.

TURKEY

Hybrid Gewehr-Karabiner *98*

Small numbers of these rifles were delivered to the Turkish Army in 1918, probably in an attempt to boost morale. Some seem to have been made from damaged full-length *Gewehr* 98, with the barrels and fore-ends considerably shortened, but others were newly made. A typical example – serial number 3571a, in the Ministry of Defence Pattern Room collection in the King's Meadow, Nottingham, factory of Royal Ordnance plc – displays the designation mark 'GEW.98' on the rear left side of the receiver, and the maker's mark 'WAFFENFABRIK', 'MAUSER A.G.', 'OBERNDORF A/N' and '1918' over the chamber in four lines. A large Turkish crescent is struck above the maker's mark. The very short barrel must have caused a very unpleasant muzzle blast, which had also been common to the original 1898-system carbine.

Bibliography

Anon., *Beute-Maschinengewehre* (Kriegsministerium, Berlin, 1916).

Anon., *Chronology of the Great War* (Greenhill Books, 1988).

Anon., *German Army Handbook April 1918* (Arms & Armour Press and Hippocrene Books, Inc., 1977).

Anon., *Text Book of Small Arms* (HMSO, 1888, 1894, 1904 and 1909 editions).

Anon., *The Training and Employment of Grenadiers* (General Headquarters, 1915).

Anon., *Waffen-Instruktion für die Artillerie und die Train-Truppe des k.k. Heeres* (Kriegsministerium, Vienna, 1882).

Ball, R.W.D., *Mauser Military Rifles of the World* (Krause Publications, 1996).

Banks, A., *A Military Atlas of the First World War* (Purnell Books Services Ltd, 1975).

Barnes, F.C., *Cartridges of the World* (DBI Books, fifth edition, 1985).

Breathed Jr, J.W. and Joseph J. Schroeder Jr, *System Mauser* (Handgun Press, 1967).

Carter, A., *German Bayonets* (Tharston Press, four volumes, 1984–94).

Datig, F.A., *The Luger Pistol (Pistole Parabellum)* (Borden Publishing Company, revised edition, 1958).

Ezell, E.C., *Handguns of the World* (Stackpole Books, 1981).

Fischer, K., *Waffen- und Schiesstechnischer Leitfaden* (R. Eisenschmidt, fifth edition, 1944).

Görtz, J. and J. Walter, *The Navy Luger* (Lyon Publishing, and Handgun Press, 1988).

Götz, H-D., *Die deutschen Militärgewehre und Maschinenpistolen 1871–1945* (Motor-Buch Verlag, 1974).

Kent, D.W., *German 7.9mm Military Ammunition 1888–1945* (privately published, 1973).

Kenyon Jr, C., *Lugers at Random* (Handgun Press, revised edition, 1990).

Korn, R.H., *Mauser-Gewehre und Mauser-Patente* (Ecksteins Biographischem Verlag, 1908).

Kromar, K. von, *Repetier und Handfeuer Waffen der Systeme Ferdinand Ritter von Mannlicher* (L.W. Seidel & Sohn, 1900).

Kropatschek, A.R. von, (ed.) *Handbuch für die kais. kön. Artillerie* (Kriegsministerium, Vienna, 1873).

Liddell Hart, Captain Sir B., *History of the First World War* (Cassell & Company, 1970).

Longstaff, Major F.V. and A. Hilliard Atteridge, *The Book of the Machine Gun* (Hugh Rees Ltd, 1917).

Lucas, J.S., *Austro-Hungarian Infantry 1914–1918* (Almark Publications, 1973).

Macdonald, L., *1914* (Michael Joseph, 1987).

– *1914–1918. Voices & Images of the Great War* (Michael Joseph, 1988).

Markham, G., *Guns of the Empire* (Arms & Armour Press, 1990).

– *Guns of the Reich* (Arms & Armour Press, 1989).

Musgrave, D.D. and Smith H.O., *German Machineguns* (MDR Associates, 1971).

Olson, L., *Mauser Bolt-Action Rifles* (F. Brownell & Sons, third edition, 1976).

Reckendorf, H., *Die Handwaffen der Koeniglich Preussischen und der Kaiserlichen Marine* (published privately, 1983).

– *Die Militär-Faustfeuerwaffen des Königreiches Preussen und des Deutschen Reiches* (published privately, 1978).

Schott, J., *Grundriss der Waffenlehre* (Eduard Zernin, 1876).

Schmidt, R., *Die Handfeuerwaffen* (B. Schwabe, two volumes, 1875–78).

Seel, Ing. (Grad.) W., 'Türken-Mauser. Mauser-Gewehre unter dem Halbmond', in the *Deutsches Waffen-Journal* (Journal-Verlag Schwend, June 1981–January 1982).

Smith, W.H.B., *Book of Pistols and Revolvers* (Stackpole Books, seventh edition, 1968).

Still, J., *Imperial Lugers and Their Accessories* (published privately, 1991).

– *The Pistols of Germany and Its Allies in Two World Wars* (published privately, 1982).

Walter, J., *German Military Handguns, 1879–1918* (Arms & Armour Press, 1982).

– (ed.) *Guns of the First World War* (Greenhill Books, 1988).

– *Luger* (Arms & Armour Press, 1975).

– *Rifles of the World* (DBI Books Division of Krause Publishing, second edition, 1998).

– *The German Bayonet* (Arms & Armour Press, 1976).

– *The German Rifle* (Arms & Armour Press, 1979).

– *The Luger Book* (Arms & Armour Press and Sterling Publishing Company, 1986).

– *The Luger Story* (Greenhill Books and Stackpole Books, 1995).

Woodward, D., *Armies of the World 1854–1914* (Sidgwick & Jackson Ltd, 1978).

Index